Nanostructuring of Metals by Severe Plastic Deformation

Payal

Table of Contents

Table of Contents

CHAPTER 1

INTRODUCTION

Aluminium alloys are considered important engineering materials because they provide a desirable combination of mechanical performance and physical characteristics. Their widespread use is mainly due to properties such as low weight with high strength, ease of fabrication, satisfactory ductility, and good corrosion resistance. Because of these advantageous characteristics, aluminium alloys have become indispensable in numerous engineering applications, especially in the automotive, aerospace, structural, and construction industries (Valievv et al., 1986; Kashyap & Chandrashekar, 2001). Conventionally, the strengthening of aluminium alloys is achieved through several mechanisms, including strain hardening, precipitation hardening, and solid-solution strengthening (SSS) (Mohanty & Gruzleski, 1995). Among these approaches, precipitation hardening is widely observed as one of the greatest active and economically viable approaches for producing high-strength aluminium alloy components. This process involves the creation of fine and homogeneously distributed precipitates within the aluminium background, which act as blockades to dislocation gesture. As a result, the resistance to plastic deformation increases, leading to a significant upgrading in the mechanical strength of stage-hardenable aluminium compounds. The effectiveness of precipitation hardening largely depends on the size, the size, distribution, and quantity of precipitates within the alloy formed during the ageing process. A finer and more uniform distribution of precipitates provides greater obstruction to dislocation movement, thereby enhancing the overall strength and performance of the alloy (Islamgaliev et al., 2017).

Precipitation hardening is achieved through natural or artificial ageing following solution heat treatment. At elevated temperatures, the solubility of solute elements in aluminium increases, whereas upon cooling the reduced solubility promotes the nucleation of nanoclusters and metastable precipitate phases, which eventually transform into stable precipitates. The scope, distribution, capacity fraction, and nature of these triggers have a major stimulus on the strengthening behaviour of age hardenable aluminium alloys. The kinetics of ageing strongly influences the precipitation process, as variations in temperature and ageing time control the nucleation and growth of strengthening phases. For specimen, in Al–Cu grounded alloys such as AA2014, the precipitation sequence typically follows the establishment of GP zones, followed by θ″ and θ′ metastable phases, before finally transforming into the equilibrium θ phase. Consequently, AA2014 aluminium alloy, a precipitation hardenable Al–Cu alloy, is commonly employed in load-bearing applications components in aerospace and automotive applications because it offers high strength along with reliable fatigue performance (Easton et al., 2011).

The growing demand for advanced materials that possess excellent properties, especially in applications involving the aerospace and automotive industries, has led to significant scientific effort being put into micro-structured materials. Nano structured (NS) and Ultra-Fine Grained (UFG) materials have received particular attention in this area because of their excellent mechanical characteristics. The enhancement of grain extent to the submicron or nano-meter gauge significantly enhances the strength and hardness of metallic materials through grain boundary strengthening, commonly defined by the Hall Petch relationship. Furthermore, materials with ultrafine and nano sized grains often demonstrate improved fatigue resistance and wear performance related with their coarse-grained matching part. Among the commonly used 'severe plastic deformation' (SPD) techniques are 'equal channel angular pressing' (ECAP), 'high-pressure torsion' (HPT), and 'accumulative roll bonding' (ARB), have been widely employed to produce NS and UFG structures in metallic alloys. These advanced microstructures provide a promising pathway for achieving a superior simultaneous improvement in strength, ductility, and structural stability in engineering materials. ("What is Aluminum,," n.d.; Davis, 2001). However, producing bulk materials with stable nano/ultrafine grain morphologies suitable for structural applications remains challenging. Severe Plastic Deformation (SPD) techniques provide a promising route for achieving substantial grain refinement through the imposition of very high plastic strains without altering the macro-dimensions of the workpiece (Wanhill & Bray, 2014; Hall, 1951; Petch, 1953). 'Severe plastic deformation techniques', including 'high-pressure torsion' (HPT) ("What is Aluminum,," n.d.) and 'equal-channel angular pressing' (ECAP) (Sharma & Kumar, 2017), multiaxial forging (MAF) (Gairola et al., 2019; Prasad, n.d.), 'cyclic extrusion and compression' (CEC) (Betsofen et al., 2016), hydrostatic extrusion (HE) (Abd El-Aty et al., 2018), twist extrusion (TE) (Langford, 1969; Sevillano, 1980), processes such as 'accumulative roll bonding' (ARB) (Verlinden, n.d.) and 'repetitive corrugation and straightening' (RCS) (Sharma & Kumar, 2017), 'friction-stir processing' (FSP) (Kumar et al., 2015; Singh et al., 2013; Joshi et al., 2016), and cryo-/cold-rolling (Valiev et al., 2000; Lavernia et al., 1990; Prasad et al., 2003) have successfully produced UFG materials with minimal dimensional change.

The key parameters influencing homogeneous grain refinement in SPD processed alloys are deformation strain, processing route, and temperature. During severe deformation, high dislocation densities are generated, which rearrange into sub grains bounded by dense dislocation walls, eventually transforming into high-angle grain boundaries (HAGBs) via continuous dynamic recrystallization (Gairola et al., 2019; Giummarra et al., 2007). However, in high and medium stacking-fault energy (SFE) metals such as aluminium and copper, dynamic recovery (DRV) facilitated by cross-slip and climb can limit dislocation accumulation and restrict grain refinement below ~100–200 nm (Budinski, 2010).

Processing at cryogenic temperatures (e.g., liquid nitrogen temperature) suppresses DRV, promotes dislocation accumulation, and enables ultrafine/nanograin formation in high-SFE metals (Babel et al., n.d.; Peel et al., 1984). Deformation at such low temperatures also increases strain localization and shear-band formation, depending on the alloying elements and initial microstructure. According to the Hall–Petch relationship (Hall, 1951; Petch, 1953), decreasing grain size increases yield strength, and in UFG materials, additional strengthening arises from nano-precipitates, dense dislocations, solute atoms, sub-grain structures, and nano-twins (Valiev et al., 2000; Ma, 2011). The combined effect of these strengthening mechanisms leads to a significant improvement in the overall mechanical performance of ultrafine-grained (UFG) alloys.

A major challenge in SPD-processed materials is achieving high strength without compromising ductility. Since SPD often reduces strain-hardening capacity, ductility losses are common. Several approaches such as bimodal grain-size distribution, grain-boundary character modification, and nano-precipitate engineering have been introduced to improve tensile elongation in UFG metals (Polmear, 2006; Rioja, 1998). When it comes to age hardenable aluminium alloys, this process is particularly effective as it will allow the precipitation of fine second phase particles which interfere with the movement of dislocations and hence restores ductility.

Apart from strength and ductility, one other important property which should be considered is rupture toughness. Unlike simple tensile testing, fracture-toughness evaluation involves high triaxial stress states and provides deeper insights into crack initiation and propagation behaviour (Giummarra, 2008; Humphreys, 2004). Therefore, a comprehensive study of fracture behaviour is essential, especially for SPD processed bulk UFG Al alloys.

The present work focuses on producing UFG AA2099 via MAF+CR evaluating their effects on microstructure and mechanical behaviour. A novel method to achieve high fracture strength in AA2099 has been proposed based on the current investigations.

Among all metals, aluminium is the most plentiful occurring in the Earth's crust, where it makes up more than 8% by mass and stands as ranking third in abundance after oxygen and silicon (Kashyap & Chandrashekar, 2001). Despite being discovered in 1824, commercial-scale extraction was not possible until the late 19th century. Today, more than 300 aluminium-containing minerals and compounds are known. Owing to its low-density lightweight in nature (2.7 g/cm^3, around one-third the density of steel) with strong corrosion resistance, excellent formability, and favourable strength-to-weight ratio, aluminium and its alloys are extensively used in household goods, transport, precision engineering, structural components, and aerospace systems.

Aluminium naturally forms a thin, adherent, transparent oxide film that protects it from further oxidation. This passive Al_2O_3 layer rapidly reforms when damaged, which makes aluminium inherently resistant to corrosion. Aluminium can be alloyed with several elements, among which copper is particularly important, giving rise to the '2xxx series' (Al–Cu alloys) that are heat correctable. After appropriate solution treatment and ageing, these alloys can achieve strength levels comparable to or even higher than those of low-carbon steels. The Al–Cu–Li alloy system, where lithium acts as a key alloying element, provides additional advantages such as reduced density and increased elastic modulus. Commercial Al–Li alloys generally contain up to about 2.45 wt.% Li. (Wanhill & Bray, 2014).

The evolution of "Al-Li alloys" began from the initial development of AA2020 (Al-Cu Li) alloys in the 1960s, followed by AA1420 (Al-Mg-Li) alloys created in the Soviet Union in the 1980s. In this case, the development of aluminum-lithium alloys has passed three generations, with major improvements being made in terms of strength, toughness, and fatigue properties (Prasad, n.d.; Betsofen et al., 2016). Moreover, the application of Rapid Solidification Technology (RST), which was introduced in the 1980s, led to even greater improvements in the properties of Al-Li alloys (Abd El-Aty et al., 2018).

Grain refinement was recognized as an important aspect that can improve the mechanical properties of metals through the work of Edgar C. Bain Hall and Norman J. Petch in their early investigations. Thermomechanical treatments used by traditional methods only generate grain sizes from several micrometres and can no longer provide improvements to mechanical properties beyond that range. However, modern processing techniques like vapor deposition, 'high-energy ball milling', rapid solidification, and several 'severe plastic deformation' (SPD) techniques have the ability to create submicron or nano-grain structures (Langford, 1969; Sevillano, 1980; Verlinden, n.d.).

The materials produced using the SPD route tend to have excellent mechanical properties, including great strength, superior fatigue properties, high toughness, and even superplastic behavior in certain cases at lower temperatures (Valiev et al., 2000).

Of the various SPD techniques, one that has gained much interest is multiaxial forging (MAF). This technique relies on the fact that the material is compressed repeatedly with 90° orientation changes in each pass. This technique results in the production of a uniform ultrafine-grained or nanocrystalline microstructure without changing the dimensions of the workpiece. Another promising technique is precipitation process, as variations in temperature and ageing time control the nucleation and growth of strengthening phases. For specimen, in Al–Cu grounded alloys such as AA2014, the precipitation sequence typically follows the establishment of GP zones, followed by θ″ and θ′ metastable phases, before finally transforming into the equilibrium θ phase. Consequently, AA2014 aluminium alloy, a precipitation hardenable Al–Cu alloy, is commonly employed in load-bearing applications components in aerospace and automotive applications because it offers high strength along with reliable fatigue performance (Easton et al., 2011).

The growing demand for advanced materials that possess excellent properties, especially in applications involving the aerospace and automotive industries, has led to significant scientific effort being put into micro-structured materials. Nano structured (NS) and Ultra-Fine Grained (UFG) materials have received particular attention in this area because of their excellent mechanical characteristics. The enhancement of grain extent to the submicron or nano-meter gauge significantly enhances the strength and hardness of metallic materials through grain boundary strengthening, commonly defined by the Hall Petch relationship. Furthermore, materials with ultrafine and nano sized grains often demonstrate improved fatigue resistance and wear performance related with their coarse-grained matching part. Among the commonly used 'severe plastic deformation' (SPD) techniques are 'equal channel angular pressing' (ECAP), 'high-pressure torsion' (HPT), and 'accumulative roll bonding' (ARB), have been widely employed to produce NS and UFG structures in metallic alloys. These advanced microstructures provide a promising pathway for achieving a superior simultaneous improvement in strength, ductility, and structural stability in engineering materials. ("What is Aluminum,," n.d.; Davis, 2001). However, producing bulk materials with stable nano/ultrafine grain morphologies suitable for structural applications remains challenging. Severe Plastic Deformation (SPD) techniques provide a promising route for achieving substantial grain refinement through the imposition of very high plastic strains without altering the macro-dimensions of the workpiece (Wanhill & Bray, 2014; Hall, 1951; Petch, 1953). 'Severe plastic deformation techniques', including 'high-pressure torsion' (HPT) ("What is Aluminum,," n.d.) and 'equal-channel angular pressing' (ECAP) (Sharma & Kumar, 2017), multiaxial forging (MAF) (Gairola et al., 2019; Prasad, n.d.), 'cyclic extrusion and compression' (CEC) (Betsofen et al., 2016), hydrostatic extrusion (HE) (Abd El-Aty et al., 2018), twist extrusion (TE) (Langford, 1969; Sevillano, 1980), processes such as 'accumulative roll bonding' (ARB) (Verlinden, n.d.) and 'repetitive corrugation and straightening' (RCS) (Sharma & Kumar, 2017), 'friction-stir processing' (FSP) (Kumar et al., 2015; Singh et al., 2013; Joshi et al., 2016), and cryo-/cold-rolling (Valiev et al., 2000; Lavernia et al., 1990; Prasad et al., 2003) have successfully produced UFG materials with minimal dimensional change.

The key parameters influencing homogeneous grain refinement in SPD processed alloys are deformation strain, processing route, and temperature. During severe deformation, high dislocation densities are generated, which rearrange into sub grains bounded by dense dislocation walls, eventually transforming into high-angle grain boundaries (HAGBs) via continuous dynamic recrystallization (Gairola et al., 2019; Giummarra et al., 2007). However, in high and medium stacking-fault energy (SFE) metals such as aluminium and copper, dynamic recovery (DRV) facilitated by cross-slip and climb can limit dislocation accumulation and restrict grain refinement below ~100–200 nm (Budinski, 2010).

Processing at cryogenic temperatures (e.g., liquid nitrogen temperature) suppresses DRV, promotes dislocation accumulation, and enables ultrafine/nanograin formation in high-SFE metals (Babel et al., n.d.; Peel et al., 1984). Deformation at such low temperatures also increases strain localization and shear-band formation, depending on the alloying elements and initial microstructure. According to the Hall–Petch relationship (Hall, 1951; Petch, 1953), decreasing grain size increases yield strength, and in UFG materials, additional strengthening arises from nano-precipitates, dense dislocations, solute atoms, sub-grain structures, and nano-twins (Valiev et al., 2000; Ma, 2011). The combined effect of these strengthening mechanisms leads to a significant improvement in the overall mechanical performance of ultrafine-grained (UFG) alloys.

A major challenge in SPD-processed materials is achieving high strength without compromising ductility. Since SPD often reduces strain-hardening capacity, ductility losses are common. Several approaches such as bimodal grain-size distribution, grain-boundary character modification, and nano-precipitate engineering have been introduced to improve tensile elongation in UFG metals (Polmear, 2006; Rioja, 1998). When it comes to age hardenable aluminium alloys, this process is particularly effective as it will allow the precipitation of fine second phase particles which interfere with the movement of dislocations and hence restores ductility.

Apart from strength and ductility, one other important property which should be considered is rupture toughness. Unlike simple tensile testing, fracture-toughness evaluation involves high triaxial stress states and provides deeper insights into crack initiation and propagation behaviour (Giummarra, 2008; Humphreys, 2004). Therefore, a comprehensive study of fracture behaviour is essential, especially for SPD processed bulk UFG Al alloys.

The present work focuses on producing UFG AA2099 via MAF+CR evaluating their effects on microstructure and mechanical behaviour. A novel method to achieve high fracture strength in AA2099 has been proposed based on the current investigations.

Among all metals, aluminium is the most plentiful occurring in the Earth's crust, where it makes up more than 8% by mass and stands as ranking third in abundance after oxygen and silicon (Kashyap & Chandrashekar, 2001). Despite being discovered in 1824, commercial-scale extraction was not possible until the late 19th century. Today, more than 300 aluminium-containing minerals and compounds are known. Owing to its low-density lightweight in nature (2.7 g/cm^3, around one-third the density of steel) with strong corrosion resistance, excellent formability, and favourable strength-to-weight ratio, aluminium and its alloys are extensively used in household goods, transport, precision engineering, structural components, and aerospace systems.

Aluminium naturally forms a thin, adherent, transparent oxide film that protects it from further oxidation. This passive Al_2O_3 layer rapidly reforms when damaged, which makes aluminium inherently resistant to corrosion. Aluminium can be alloyed with several elements, among which copper is particularly important, giving rise to the '2xxx series' (Al–Cu alloys) that are heat correctable. After appropriate solution treatment and ageing, these alloys can achieve strength levels comparable to or even higher than those of low-carbon steels. The Al–Cu–Li alloy system, where lithium acts as a key alloying element, provides additional advantages such as reduced density and increased elastic modulus. Commercial Al–Li alloys generally contain up to about 2.45 wt.% Li. (Wanhill & Bray, 2014).

The evolution of "Al-Li alloys" began from the initial development of AA2020 (Al-Cu Li) alloys in the 1960s, followed by AA1420 (Al-Mg-Li) alloys created in the Soviet Union in the 1980s. In this case, the development of aluminum-lithium alloys has passed three generations, with major improvements being made in terms of strength, toughness, and fatigue properties (Prasad, n.d.; Betsofen et al., 2016). Moreover, the application of Rapid Solidification Technology (RST), which was introduced in the 1980s, led to even greater improvements in the properties of Al-Li alloys (Abd El-Aty et al., 2018).

Grain refinement was recognized as an important aspect that can improve the mechanical properties of metals through the work of Edgar C. Bain Hall and Norman J. Petch in their early investigations. Thermomechanical treatments used by traditional methods only generate grain sizes from several micrometres and can no longer provide improvements to mechanical properties beyond that range. However, modern processing techniques like vapor deposition, 'high-energy ball milling', rapid solidification, and several 'severe plastic deformation' (SPD) techniques have the ability to create submicron or nano-grain structures (Langford, 1969; Sevillano, 1980; Verlinden, n.d.).

The materials produced using the SPD route tend to have excellent mechanical properties, including great strength, superior fatigue properties, high toughness, and even superplastic behavior in certain cases at lower temperatures (Valiev et al., 2000).

Of the various SPD techniques, one that has gained much interest is multiaxial forging (MAF). This technique relies on the fact that the material is compressed repeatedly with 90° orientation changes in each pass. This technique results in the production of a uniform ultrafine-grained or nanocrystalline microstructure without changing the dimensions of the workpiece. Another promising technique is cryorolling (CR), which is carried out at cryogenic temperatures. Processing under such conditions effectively suppresses dynamic recovery and dislocation annihilation, thereby enabling the accumulation of a high dislocation density within the material. As a consequence, cryorolling facilitates the development of ultra-acceptable grains and significantly enhances the strength of the material even at relatively low levels of applied strain (Kumar et al., 2015; Singh et al., 2013; Joshi et al., 2016).

In response to the growing demand for lightweight materials with high strength for aerospace and structural applications, the current study explores the development of an ultra-small-grained Al–Li 2099 alloy using a combined severe plastic deformation route involving multiaxial forging (MAF) and cryorolling (CR), afterwards post annealing treatment. The work aims to understand the deformation mechanisms that promote grain refinement during the combined SPD processing. Particular attention is given to the evolution of microstructure and the changes occurring under successive deformation steps. In addition, the influence of post-annealing on precipitate formation and the associated strengthening mechanisms is examined. The mechanical performance of the processed alloy is appraised in relations of tensile strength, hardness, and fracture toughness. A thorough analysis of the relationship among deformation processing, grain refinement, precipitation phenomena, and mechanical properties is essential for designing and developing advanced 'aluminium alloys' for future applications.

Fracture toughness is important because it helps evaluate crack initiation and crack propagation behaviour under complex stress conditions. Therefore, fracture toughness is essential for judging the structural reliability and practical usability of SPD-processed UFG aluminium alloys.

1.1 Aluminium Lithium Alloy 2099

Lithium is the lightest metal element, having a density of about 0.534 g/cm^3, almost five times less than that of aluminium. The use of such an element in aluminium alloys for reducing density is a subject of significant interest, especially for applications where weight reduction is considered crucial in design. In recent decades, aluminium-lithium (Al-Li) alloys have been increasingly replacing many conventional aluminium alloys in aerospace and military systems due to their improved density and mechanical properties.

The inclusion of lithium is very significant when it comes to changing the physical and mechanical properties of the aluminum alloys. In many studies conducted on these alloys, it has been noted that when about 1 wt% of lithium is introduced into them, their density reduces by around 3%, and at the same time, their elastic modulus increases by around 6%. Due to these two beneficial changes, the specific mechanical properties of such alloys are much better than those of ordinary aluminum alloys.

There have been significant advances made in the development of new 'Al–Li alloys' in the latter part of the twentieth century. Specifically, the contributions made by John R. Pickens et al in 1989 in Martin Marietta Corporation helped in the development of third-peer group Al-Li alloys. This work paved the way towards the forecast of weldable, less compact aluminum alloys that were capable of storing fuel in cryogenic conditions and also in aerospace launch vehicles. In such an alloy, the concentration of lithium is usually kept below about 2 wt%. The alloy AA2219 was frequently chosen as the base metal owing to its excellent mechanical properties at cryogenic conditions.

Previous research work carried out by John Peel et al. also provided valuable contributions to the study of how lithium content affects the physical behavior of aluminium alloys. In their research, they established an empirical correlation that explains the decrease in density that occurs when lithium is added into aluminium alloys. The empirical correlation has been helpful in providing references for future research on the density of Al-Li alloys.

$$\rho(g / cm3) = 2.71 + 0.024Cu + 0.018Zn + 0.022Mn - 0.01Mg - 0.004Si - 0.079Li \text{---}(1)$$

High intensity of Brass-type texture and pronounced in-plane anisotropy have been predominantly reported in second-generation rolled Al–Li alloy products (Rioja, 1998; Giummarra, 2008; Humphreys, 2004). Second-generation Al–Li alloys typically contained lithium contents of about 2 wt.% or slightly higher. Although these alloys offered advantages such as reduced density and increased stiffness, several limitations were also observed. These included relatively poor high-cycle fatigue performance, lower impact toughness, reduced short-transverse ductility, and a decline in fracture toughness. In addition, the alloys were susceptible to degradation in toughness when exposed to thermally unstable conditions (Prasad et al., n.d.).

To address these drawbacks, the third peer group of 'Al–Li alloys' was developed in the early 1990s. In these alloys, the lithium content was deliberately reduced to below approximately 2 wt.% in order to achieve a more favourable balance between strength, toughness, and structural reliability while retaining the benefits associated with lithium additions. Alloys such as 2076, 2065, 2055, 2060, 2050, 2199, 2099, 2397, 2297, 2198, 2196, and 2195 were specifically industrialized for aerospace and airplanes applications and are classified as third peer group Al–Li alloys. Notably, alloys 2195, 2196, 2297, 2397, 2198, 2099, 2199, 2050, 2060, and C99N have been extensively studied and optimized for space and aviation structures.

In current years, significant attention has been directed toward the progress of high-strength aluminium alloys for advanced structural applications. Among these, alloys fitting to the Al–Zn–Mg–Cu system and lithium-comprehending aluminium alloys, commonly known as Al–Li alloys, have confirmed significant enhancements in strength and structural performance (Prasad, n.d.). As a result, these materials endure to play a significant character in modern aerospace engineering, where a 'high strength-to-heaviness' ratio and enhanced stiffness are essential for achieving improved structural efficiency.

AA 2099 is a heat-treatable aluminium–lithium alloy that was initially designated as C 458. The alloy was first developed in 1997 and subsequently introduced commercially by Alcoa during the period 2003–2004 (Ma, 2011; Polmear, 2006). Earlier generations of Al–Li alloys were often characterized by pronounced anisotropy and variations in mechanical behaviour, which largely resulted from the strong deformation textures formed during thermomechanical processing (Rioja, 1998). In contrast, AA 2099 belongs to the third peer group of 'Al–Li alloys' and was specifically engineered to provide improved mechanical performance, lower density, and better fabrication characteristics compared with second-peer group alloys such as AA 2090 and AA 8090.

AA 2099 demonstrates a grouping of favourable physical and mechanical possessions that distinguish it from many conventional aluminium alloys of the 2xxx and 7xxx series. One of the key reasons for this improvement is the significantly reduced concentration of alkali metal impurities in C-458 (AA 2099) when compared with earlier Al–Li alloy compositions. As a result, the alloy exhibits higher strength, improved resistance to fatigue crack propagation, and enhanced corrosion resistance relative to traditional 2xxx and 7xxx alloys. In addition, AA 2099 provides a density reduction of approximately 5.4%, which contributes to improved specific mechanical performance (Prasad, n.d.).

This alloy was made in the hope that it could replace commonly used aluminum alloys like the 2024, 2026, 6110, 8090, 7050, 7075, and 7175 series. While providing equal or even better mechanical properties, AA 2099 possesses good fatigue characteristics, along with enhanced corrosion resistance in comparison with alloys like 7075-T6, especially in aqueous solutions.

Temper 2099-T83 falls into the category of high-strength temper and offers much better yield strength than that of traditional 2024-T42 aluminum alloy. In addition to this, a different tempering process 2099-T81 aims at improving damage tolerance but also manages to offer more strength than 2024-T42 alloy. Because of its superior combination of strength, damage tolerance, and low density, 2099 alloy can be referred to as a lightweight structural alloy.

1.1.1 Applications of Al-Li Alloys

Al-Li alloys have been applied in the manufacture of modern commercial aeroplane structures. For instance, the application of Al-Li alloys is found in the major physical systems of aeroplanes such as Boeing 787 Dreamliner, Boeing 777-9 and Airbus A350, due to their low density and relatively high stiffness, thus contributing to efficient and effective structures (Prasad, n.d.). Other than these applications, there have been many other areas in which Al-Li alloys receive significant consideration from both researchers and industries, as discussed below. Their favourable combination of low weight, high stiffness, and good mechanical performance makes them suitable for aerospace structures, aircraft fuselage components, space vehicles, and even certain automotive applications. In particular, their ability to perform reliably under both static and cyclic loading conditions enhances their suitability for critical fuselage structures in advanced aircraft design. Representative applications of Al–Li alloys include several structural and aerospace components (Prasad, n.d.):

(a) Lower wing members, fuselage members, and pressure cabin members stiffener-dominated structural designs.

(b) Fuselage structural stiffeners and frames, upper wing stringers, as well as floor beams and seat rails in the Airbus A380

(c) F-16 aircraft fuselage panels and structural bulkheads

(d) Internal sections of the fuselage and upper wing covering panels in the Airbus A380.

(e) Military vehicles and marine industry components. Weapon system components and stiffness-dominated engineering designs.

The alloy is suitable for lightweight, high-strength structural applications, especially in aerospace components such as fuselage panels, wing members, stiffeners, frames, floor beams, seat rails, and other stiffness-dominated structures, as well as selected automotive, marine, military, and space vehicle applications.

1.1.2 Effects of alloying elements on Aluminium

Copper

Aluminium-copper alloys comprising of nearly 2 to 10% Cu, that form important families of 'Al-Cu alloys' with accumulation of other elements. Al-Cu alloys in both forms either cast or wrought alloys responds to solution heat treatment and with successive aging treatment with an enhancement in overall mechanical properties while low percentage elongation. The maximum amount of strengthening is achieved when copper is added between 4 and 6% Cu, that might dependent upon the effects of additional element present in the alloy (Davis, 2001). Addition of Copper tends to intensification of strength and toughness as Cu forms precipitates i.e. Al_6CuLi_3(T2), Al_2Cu (Θ'), and Al_2CuLi (T1) (Balducci et al., 2017). Addition of Copper also raises sensitivity during quench during heat treatment. Overall, copper diminishes the resistance towards general corrosion, on the other hand plays important role in resistance to stress corrosion by enhancing it. The major consequence of copper lead to rise in the aging rate by enhancing the gradation of super capacity via nucleation of the Al_2CuMg phase (Davis, 2001).

<table>
<tr><th colspan="6">Al-Li Alloy</th></tr>
<tr><td>Al-Mg-Li</td><td>Al-Cu-Li</td><td>Al-Li-Cu-Mg</td><td colspan="3">Al-Cu-Li-Mg</td></tr>
<tr><td></td><td></td><td></td><td>Case High Mg Medium Cu & Li Contant</td><td>Case High Li medium Cu & Mg Contant</td><td>Case High Cu Low Mg and Li Contant</td></tr>
<tr><td>(δ')
Ex.
AA1420
AA1421
AA1423</td><td>(δ', T1, Θ')
Ex.
AA2020
AA1450
AA1460</td><td>(δ', T1, S')
Ex.
AA8090
AA8091
AA1440</td><td>(δ', T1, S')
Ex
AA2091
AA1430</td><td>(δ', Θ', S')
Ex
AA2090
AA2099
AA2196
AA2197
AA2198</td><td>(T1, δ', Θ')
Ex
AA2090
AA2099
AA2196
AA2197
AA2198</td></tr>
</table>

Figure 1.2 Precipitates phases in (Al-Li-Cu) and (Al-Li-Cu-Mg) alloys (Balducci et al., 2017)

1.1.1.2 Lithium

Lithium is added to the Aluminium Alloys nearly 0.01% - 3%. This major addition of lithium in aluminium alloy significantly lead to enhancement in the oxidation rate aluminium in molten form. The current attention is focussed on the Al Cu-Li-Mg alloys, that are heat treatable alloys results in the high strength as compared to those existing 7xxx alloys. Although consequences by the addition of lithium shows lessening in density and increment in the modulus. This enhancement is caused due to the formation of a greater volume fraction of coherent, well-ordered Al3Li (δ') precipitate which highly influence the elastic modulus and fatigue crack growth resistance by increasing these properties at transitional stages of stress intensity (Davis, 2001).T2 (Al6CuLi3) and T1 (Al2CuLi) precipitates are also formed when lithium is added to Al Cu alloys which also results in enhancing toughness and overall mechanical strength of aluminium alloy (Balducci et al., 2017).

1.1.1.3 Zinc

Zinc is added in AA 2099 up to 0.4% - 1.0%. The addition of copper with zinc 4 1 1 28 together with addition of minor but vital volumes of chromium and manganese causes in the high strength Al base compounds. In Al-Cu-Li compounds system, zinc acts as a regulatory the aging process. Zinc addition to the Al-Cu-Li alloy consequences in the increment of decomposition resistance of the compound and it is predictable to be in solid solution. Although, there is no exact evidences of this in the microstructure of the compound (Balducci et al., 2017).

1.1.1.4 Magnesium

Magnesium is added in the Aluminium alloy containing copper and lithium (Al-Cu–Li system) in 0.1% - 0.5% amount. Magnesium precipitates has been assumed to be in substitute with lithium which forms T1phase for the formation of Al2(Cu, Li-Mg) precipitates. When Magnesium is added to Al-Li alloy is also responsible for the strengthening enhancement. Development of θ′ (Al_2Cu), S′ (Al_2CuMg), and σ ($Al_5Cu_6Mg_2$) precipitate phases within the alloy matrix. in Al-Cu-Li alloy is due to addition of Mg (Balducci et al., 2017).

1.1.1.5 Manganese

Manganese is added in AA 2099 in 0.1% - 0.5% by weight. Manganese forms Al20CuMn3precipitate when mixed in such a little amount in Al-Cu-Li alloy. It is of rod-like moulded dispersoids, It is essential for controlling texture evolution, inhibiting grain boundary motion, and enabling sub-grain rearrangement. and also plays a vital role to inhibit recrystallization and thus helps in enhancing the fracture toughness. However, Mn in the form of dispersed precipitate is highly effective in decelerating the recovery and stops grain growth (Balducci et al., 2017).

1.1.1.6 Zirconium

Zirconium is added in AA 2099 up to 0.05% - 0.1% amount. Generally, as similar to Mn, Zr is also addition in Al-Cu-Li alloy in low quantities which leads to formation of Al3Zr precipitate which is spherical in size. Zirconium addition also results in development of dispersoids that helps for controlling the texture, pin grain and sub-grain movement and similarly inhibits the recrystallization of grains, causes increment in fracture toughness. The joint addition of Mn and Zr elements is assumed to yield a more even dispersion coverage. These inter metallic precipitates cause lessening in resistance to recrystallization (Balducci et al., 2017).

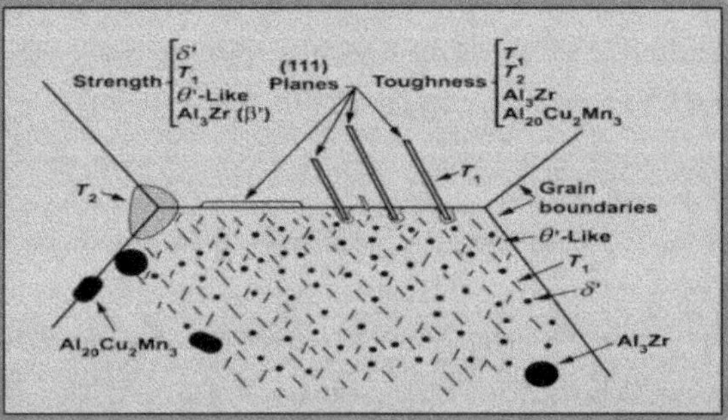

Figure 1.2 Schematic representation of the microstructural features: - Al–Cu–Li 2099 alloy, aluminium matrix. (Balducci et al., 2017)

1.2 Strengthening Mechanism in Materials

A variety of approaches are employed in materials design because the required 2 5 properties vary significantly depending on the intended application. Consequently, materials must be developed and optimized according to the specific service conditions in which they will operate. In most engineering applications, metals and their alloys are commonly used as load-bearing materials. As a result, adequate mechanical strength becomes a critical requirement for ensuring reliable performance. The need for strengthening arises because pure metals are generally relatively soft by reason of the properties of metallic attachment, which is non-directional and relatively diffuse. In addition, the presence of numerous lattice defects within the crystal structure facilitates dislocation movement, leading to plastic deformation at comparatively low stress levels. Therefore, various strengthening mechanisms are introduced to enhance the strength and mechanical stability of metallic materials. Dislocation in an important defect present in the lattice, having high mobility outstanding to the long-term of periodicity in lattices of metals and the trivial translation vectors. Any change in the behaviour of these metallic bonds or in the crystal structure for increase the strength is difficult and can't be suggested as it results in transition of ductile metallic bond into brittle solids. Although to overcome these defects and to enhance the overall mechanical properties, fine grain particles, adding solute particles in large amount to bring new transformations in metallic phases and various other methods are frequently required for the increment in the strength. The main strengthening mechanism is mentioned below.

1.2.1 Strengthening from Grain Boundaries

The boundaries present between the adjacent grains in the polycrystalline material are the regions where the lattice is distributed for only few diametric ranges. And thus, the grain orientation and grain size highly influence the overall belongings of the substantial. These grain boundaries act as the hindrances to dislocations present in the material and therefore, these dislocations mound up at the grain boundaries. During plastic deformation, dislocations tend to accumulate and form dense arrangements along cell walls or sub-boundaries. With continued deformation, these dislocations gradually reorganize into more stable, low-energy configurations at the boundaries of the cells. The interaction and accumulation of dislocations at these interfaces restrict further dislocation movement, which may lead to a reduction in ductility. However, this phenomenon can be effectively utilized to enhance the strength of the material through sub-structure or grain-boundary strengthening. The increase in strength associated with the establishment of such stable disruption substructures is generally connected to the inverse of the cell size (D), indicating that smaller cell dimensions contribute to higher strengthening.This relationship between the YS and size of grain size was proposed by Hall-Patch

$$\sigma_y = \sigma_0 + kd^{-1/2} - - - - - - - - -(2)$$

where k is known as the strengthening constant, d denotes the mean grain size, σo is the 'frictional stress' and σy is the 'yield stress'.

1.2.2 Strain Aging Treatment

Strain aging is generally a type of behaviour in material linked with the yield point phenomenon, according to which there is certain increment in strength while reduction in ductility which occurs due to the cold working in material followed by heat treatment at a relatively low temperature.

Solid Solution Strengthening or Hardening

This technique is possibly one of the oldest and was magnificently practised by experimenters in old time for the development of ornaments made of bronze, brass and gold. In solid solution strengthening, the atoms of the base metal may be swapped by solute atoms, or the solute atoms may inhabit the interstitial positions within the crystal lattice. In both situations, the presence of these solute atoms distorts the lattice structure and restricts dislocation movement, thereby increasing the strength of the material. In this process whenever a bigger substitutional solute atom enforces a compressive stress whereas the minor interstitial atom is the cause of tensile stresses present in the lattice. These interstitial atoms when larger in size than the interstitial space present in the lattice which is occupied by them, results generation of compressive stress field. Whereas, whenever these solute particles are comparable in size to the solvent particles occupies the lattice point present in the crystal lattice and are known as substitutional solid solution. This process generally leads in the enhancement in the YS and fracture toughness. Also, this mechanism affects the whole stress-strain curve. Due to this frictional resistance to the dislocation movement is more influenced rather than static locking of these dislocations.

1.2.3 Strengthening via Fine Particles

A common source of strengthening in materials is the addition of small-scale particles that are distributed in ductile matrix. There are two ways of this kind of addition. Firstly, dispersion hardening which hard particles are made to mix with the metallic matrix of powder and further consolidated and processed via powder metallurgy technique. Second is precipitation hardening also known as age hardening Alloying the material with a precipitate to a subsequent phase is possibly the furthermost active and effective strengthening path. The 2nd phase particles can be merged by precipitation of unbalanced solutes from the resolution. The subdivision size, volume by section and the nature of these 2nd phase alloys are considered as a strength of boundary between various disarticulations and these second phase particles. Formation of these fine precipitates lead to enhancing the strength and hardness of some alloys. At elevated temperatures the solute tends to have growing solubility (e.g. Al-Cu) for the occurrence of precipitation. This precipitation process includes the heat treatment process i.e. Solution treatment or heating to one phase region trailed by quenching i.e. rapid cooling to obtain a 'super saturated solid solution' (SSS) and further aging at a lower temperature. This occurs due to the coherency or matching of atoms between the 2nd phase of the particles and metallic matrix.

1.2.4 Strain hardening or Work Hardening

Strain hardening is referred to the process in which there is rapid increase in the strength and hardness by the plastic deformation. Strain hardening is also referred as cold working because the deformation in the material takes place at ambient temperature. Due to which there is increment in the degree of cold working (% area reduction) as the extent of strain hardening. The cold working process deforms the metal in a particular direction, leads to elongation in grains in the direction of cold working. Dislocations are the prime factor of strain hardening in the materials. Interaction of the stress field of these dislocations is the major cause of strain hardening. Due to the generation of dislocation during cold working process cause increase in dislocation density. And with high density of dislocations obstructs the movement due to interactions of strain field dislocation. Hence, there is an increment in the strength with lessening the ductility. The relationship between these dislocations and YS is given by.

$$\sigma_0 = \sigma_1 + \alpha G b \sqrt{\rho} - - - - - - - - - - -(3)$$

Where G is the 'shear modulus', ρis the dislocation density, b is 'Burger vector', and α is a constant.

1.2.5 Two-Phase Strengthening

A significant class of engineering alloys derives strength from the presence of more than one phase within the microstructure. In such systems, strengthening arises through mechanisms associated with both precipitation hardening and grain-boundary effects. As a result, the microstructure typically consists of an intermediate mixture of two or more distinct phases distributed within the matrix, which collectively contribute to the overall mechanical strength of the material. Microstructures lies in this category comprises of eutectic, eutectoid, and various other transformation in its structures. Such kind of alloys originates their strength from fine structure formed after mixing. A notable characteristic of alloys strengthened through a two-phase microstructure is their ability to retain considerable ductility and energy-absorption capacity. This behavior can be observed even when one of the constituent phases is relatively brittle, as the interaction between the phases helps distribute deformation more effectively.

1.2.6 Resistance to Disruption-Tempted Flow

The plastic movement of a material through dislocation motion is largely governed by several intrinsic characteristics of the multitude material, including its 'elastic modulus' (E), 'diffusion coefficient' (D), and 'crystal structure'. The energy connected with a disruption is proportional to the 'elastic modulus' and the square of the 'Burgers vector' (b). At raised up temperatures, the mobility of dislocations is also influenced by atomic diffusion within the lattice, which is described by the diffusion coefficient of the material. Since parameters such as E, b, and, to a considerable extent, D are inherent properties of a material, they cannot be significantly modified through conventional design strategies. Consequently, strengthening approaches focus on reducing dislocation mobility by introducing obstacles within the lattice. Such obstacles may include additional disruptions, sub-grain boundaries, grain boundaries, solute atoms, and precipitate particles. The interaction of moving dislocations with these features impedes their motion and thereby contributes to a development in the mechanical belongings of the material.

CHAPTER 2

NANO STRUCTURING OF METALS

Herbert Gleiter was the first who had given the concept more than 20 years ago for the development of the nano crystalline materials i.e. ultrafine-grain materials having grain size less than 100 nm possessing special properties. Later then, there had been rapid growth in the field of nano structured materials, owing to incredible interest in this topic associated to scientific and technical importance. The innovative idea of Gleiter's was that the nano crystalline materials comprises of a very large portion of grain boundaries having special atomic structure. As a result, many unusual properties had been found in nano-grain materials. As like in mechanical properties, individual could assume excessive strength and fracture toughness, high fatigue life and high resistance to wear. Nano structuring had been

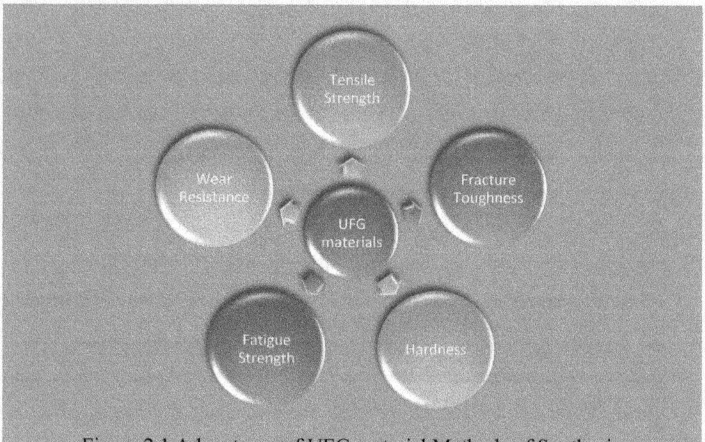

Figure 2.1 Advantages of UFG material Methods of Synthesis

expected to a revolutionary usage of nano grained materials for numerous efficient and structural applications. But these exciting prospects were put in threat as several investigations presented that although nano crystalline materials consist of very high strength and hardness. An UFG material and their alloys. The unique microstructure of UFG material makes it advantageous over these conventional alloys [40].

2.1 Introduction to Various SPD Techniques

The SPD processes are well-distinct as the 'metal-forming processes' which introduces an extreme-large plastic strain into a bulk metal/alloy to produce the UFG metal/alloys. The primary aim of SPD processes is to fabricate the light-weight and high-strength parts with ecological harmony (Prasad et al., n.d.). The grain-size of polycrystalline metals plays an important role, and affects its characteristics like fracture-toughness, fatigue, ductility and strength. The literature is well available on physical and mechanical behaviour of conventional material having grain size of several micrometers. UFG materials shows the exceptional mechanical, chemical and physical properties. The grain size of UFG microstructure ranges from 100-1000 nm and of nano materials ranges from 1-100 nm ("What is Aluminum,," n.d.; Valiev et al., 2000; Abd El-Aty et al., 2018). Generally, plastic strain about less than 2.0 is imposed in conventional metal-forming techniques namely rolling, extrusion and forging. During other SPD processes such as multi-pass drawing, extrusion and rolling the diameter and thickness of processed material becomes very thin due to the imposed plastic strain about more than 2.0 and are not appropriate to be used for parts in structural applications. Various SPD techniques have been developed to achieve the huge-strain on bulk-metal/alloy without altering its shape. The popular SPD techniques are ECAP, ARB, constrained groove pressing (CGP), continuous confined strip shearing (C2S2) and TE (Sevillano, 1980; Davis, 2001; Balducci et al., 2017; Valiev, 2004). Among conventional sheet metal forming techniques, asymmetric rolling and CR process have been extensively used to achieve UFG/NC grain structure in the material. Various SPD processes for fabricating ultra- fine grains in the bulk constituents are well reported in the works. The various existing SPD techniques are as follows:

1. Equal Channel Angular Pressing (ECAP) (Sherby, 2001; Wang, 2006; Tanimura, 1980)

a. Repetitive Side Extrusion (Langdon, 2010; Bridgman, 1946)

b. Continuous Manufacturing of Bolts (Bridgman, 1943)

c. Continuous Confined Strip Shearing (C2S2) (Balducci et al., 2017; Valiev, 2006; Mohanty & Gruzleski, 1995)

d. Rotary-die ECAP (Kashyap & Chandrashekar, 2001; Kumar et al., n.d.)

e. Incremental ECAP (I – ECAP) (Romios, 2005; Samantaray, 2010)

f. Parallel channel ECAP (Bois-Brochu, 2012; Yang et al., 2017)

g. ECAP-conform (Lin & Zheng, 2016; Vander Kooi, 2000; Betsofen et al., 2016)

h. Con-shearing (Brodusch et al., 2012)

2. Extrusion based Techniques

a. Twist extrusion (Sevillano, 1980; Carrick et al., 2013; Ma et al., 2011)

b. Tube Channel Pressing (TCP) (Abd El-Aty et al., 2018)

c. Pure Shear extrusion (Easton et al., n.d.)

3. High Pressure Torsion (HPT) ("What is Aluminum,," n.d.; Tanimura, 1980; Yogesha et al., n.d.)

a. Hollow cone-HPT process (Ringeval et al., 2006)

b. Continuous High-Pressure Torsion (Aoba et al., 2017)

4. Rolling based Techniques

a. Accumulative roll Bonding (ARB) (Verlinden, n.d.; Joshi et al., 2016; Bereczki et al., 2014)

b. Wavy rolling (Kapil et al., 2019; Kapoor et al., 2013; Yogesha et al., n.d.)

c. Cryorolling (Yogesha et al., n.d.; Rangaraju et al., 2005; Panigrahi & Jayaganthan, 2011)

d. Asymmetric rolling (Gairola et al., 2019; Joshi et al., 2016)

5. Multiaxial Forging (MAF) (Prasad, n.d.; Malekjani, 2011; Pathak, 2018; Majta et al., 2014)

a. Cyclic Close-die forging (CCDF) (Salimgareeva et al., 2006)

b. Channel Die Forging (CDF) (Yogesha et al., n.d.)

6. Other Techniques

a. Repetitive-Corrugation and Straightening (RCS) (Sharma & Kumar, 2017; Ramesh et al., 2019)
b. Continuous-RCS (Sharma & Kumar, 2017; Kulczyk et al., 2007)

2.1.1 Bottom-Up Approach, Top-Down Approach

Basically, there are two approaches in engineering for the fabrication of bulk 2 UFG/NC grain structure: (i) bottom-up and (ii) top-down approach. The representation of both is exposed in Figure 2.1. In bottom-up approach atom by atom are assembled similar to the building blocks to produce the NC material. The best suitable examples for bottom-up approach are chemical/physical deposition, cryo-milling with hot isostatic pressing, ball milling with subsequent consolidation, electro- deposition, and inert gas condensation. In top-down approach, the bulk metal/alloy is refined to fabricate the UFG structure in a material. The great extent of dislocation- density is generated in the bulk material owing to the huge amount of plastic-strain enforced. The dimension and original form of the specimen remains the same in this technique which shows the novelty of this route. The various advantages of top-down approach are listed as follows:

a) Bulk NC/UFG grained structure materials can be fabricated through SPD.
b) The grain size as well as the defects introduced during casting can be removed without affecting the specimen dimension.
c) SPD processing can be done through metal forming techniques and can be easily adopted by the industry.

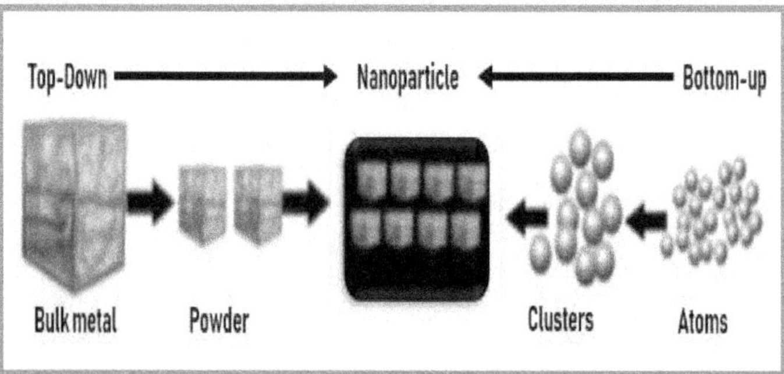

Figure 2.2 Schematic representation of top-down and bottom-up approach

2.1.2 Identical Channel Angular Pressing Technique

One of the commonly used SPD technique is ECAP process. In 1981, Segal and his co-workers was introduced the ECAP process for the production of UFG/NS (nanostructured) materials (Odeshi et al., 2019). In this process the material

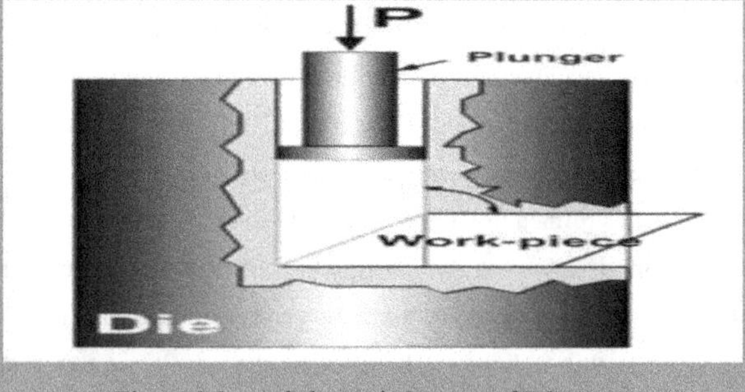

Figure 2.3 Schematic Layout of ECAP.

undergoes strong plastic- strain by simple shear to achieve the grain refinement of material. In this technique, the sample having uniform cross-section is placed in die having intersecting channel form under the minimum friction conditions. The plunger is used to press the sample. To achieve the higher value of strain the pressing may be repeated and the orientation of material blank is rotated along the length direction during every pass. The amount of strain can also be increased by applying the back pressure at the end of the sample.

This method consists of four different shear routes to make product with distinct microstructure. Figure 2.3 shows the four different processing shear routes for the bar/rod sample (Muñoz-Morris & Morris, 2010). The sample without rotation is pressed in first processing route A. The rotation at an angle of 90° in alternative route during each pass shows the second processing route BA. In third process route BC sample is rotated by an angle of 90° in clockwise/anticlockwise path during each pass. In the fourth i.e. final process route C, the sample is rotated by an angle of 180° during each pass. The other possible combinations of these processing routes are also possible but the literature available to date recommends that such combinations din not contribute supplementary enhancement in mechanical characteristics of as-pressed metal/alloys.

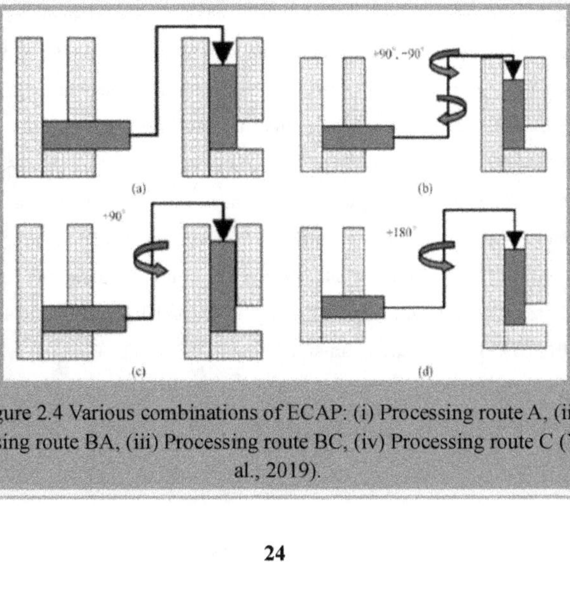

Figure 2.4 Various combinations of ECAP: (i) Processing route A, (ii) Processing route BA, (iii) Processing route BC, (iv) Processing route C (Yin et al., 2019).

This technique is a suitable because load requirement is very less and simple tool geometry resulting low tool pressure to develop new microstructure with improved strength in the material at applied strain (Li et al., 2018).

2.1.3 Multi-Directional Forging Technique

One of the novel SPD technique for fabricating the bulk UFG alloys is Multi-Directional Forging (MDF). This technique was very famous throughout the initial (½) of 1990s for getting the UFG structure in bulk metal/alloy. (Malekjani, 2011; Pathak, 2018). In multi-directional forging the mechanism responsible for grain refinement is continuous dynamic recrystallisation. Figure 2.4 represents the schematic layout of MDF process. This process involves the numerous repetitions of free forging operation along with the pulling and setting as well as the change of axis for load applied (Qin et al., 2009). In comparison to HPT and ECAP process the strain produced in this process is less homogenous. This method suits best for obtaining the nanostructured state in brittle material due to its capability of functioning at elevated temperatures. The proper combination of strain- rate of deformation along with the temperature consequences in obtaining the wanted grain- enhancement. Generally, this development is achieved for the absolute melting temperature (Tm) range of (0.1–0.5) Tm. The output of this process is large-size billets having the nanocrystalline structures. In literature it was reported that this process is used for various temperature ranges ranging from cryogenic temperature (-196° C) to elevated temperature (600° C) for microstructural refinement of various alloys such as Plain low carbon steel (Folias, 1974), titanium alloy (Malekjani, 2011), pure Aluminum (AA1100) (Gollinga et al., 2018), Cu (Qin et al., 2009), Al-Mg alloy (Ovid'ko & Sheinerman, 2012), Mg alloys (Hohenwarter & Pippan, 2010; Verma et al., 2018; Dhal et al., 2018), Al 6061 alloy (Prasad & Malakondaiah, 1992), AISI 1016 steel (Wronski & Bacroix, 2014; Wronski et al., 2013), Fe-32%Ni alloy (Kumar, 2015).

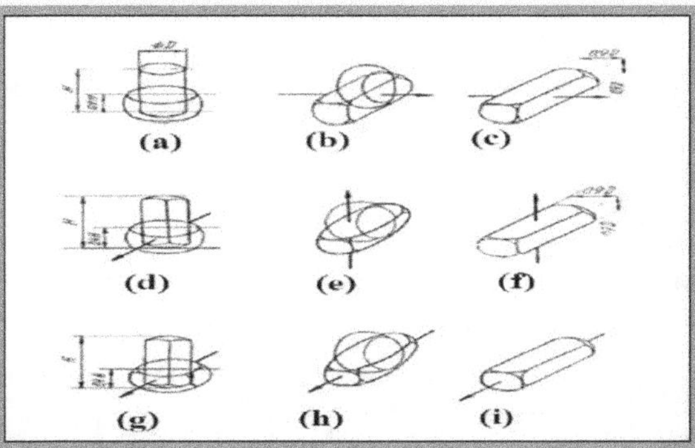

Figure 2.5 MAF process showing setting and pull-broaching along the first axis (a – c), along the second axis (d – f), and along the third axis (g – i). (,Malekjani, S. et al., 2011, Pathak, M. K. et al.,2018)

2.1.4 Accumulative Roll Bonding Technique

'ARB' technique for making UFG structure in metal/alloys was initially proposed by Saito et al. (Betsofen et al., 2016). The representation illustration of ARB technique is shown in Figure 2.5, where two roll bonded sheets were cleaned with the help of acetone and steel wire brush. The thick sheet form is then obtained by stacking two strips one above another followed by conventional rolling process to achieve ~50% thickness reduction in each pass. The processed material is then cut into two halves followed by cleaning thoroughly with wire brushing then stacking together and finally the rolling again. The repetition of this process is done, till the desired value of strain is achieved in material. The strength of the ARB processed metal/alloy critically depends on bond quality, which may be challenging to attain.

The process may be implemented at RT (area temperature) as well as at some

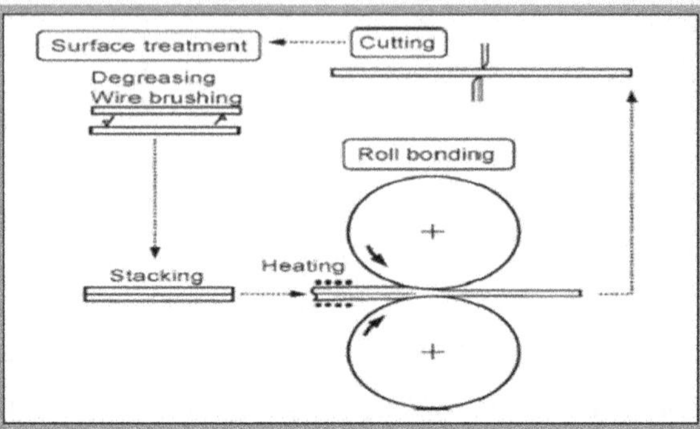

Figure 2.6 Schematic representation of ARB process (Davis, 2001).

preeminent temperature. The basic mechanism of grain refinement in metal/alloy during ARB process takes place due to the pure shear (Verlinden, n.d.; Joshi et al., 2016). UFG metals/alloys fabricated through ARB process exhibits high strength and ductility (Lynch et al., 2014). During recent years, this process has been successfully used to fabricate metallic multi-layered composites by performing roll bonding between similar/dissimilar metals and alloys (Verlinden, n.d.; MacMaster et al., 2000; Rout et al., n.d.; MacRae et al., 2018). The advantages of this process are simplicity, cheap, and easy adoptability in Figure 2.6 Schematic representation of ARB process (Davis, 2001).Figure 2.6 Schematic representation of ARB process (Davis, 2001).industrial process.

2.1.5 Cyclic Extrusion Compression Technique

Richert et al. (Hall, 1951) devised the cyclic extrusion compression (CEC) process well known for producing UFG structure in metal/alloys. Figure 2.6 represents the schematic layout of CEC technique. This process contains the cyclic-flow of metals/alloys between the blinking extrusion as well as the compression chambers. The adequate counter-pressure is provided to the samples to cause plastic-deformation by alternating extrusion and compression by touching the work piece cyclically between the two die chambers to build up desired plastic-strain. The approximate equivalent-strain (accumulated) is given by the following equation 2.1

$$\varepsilon = 4n\ln\left(\frac{D}{d}\right) \quad \ldots\ldots\ldots \quad 2.1$$

Where, 'd' denotes channel diameter, 'D' denotes chamber diameter and 'n' denotes

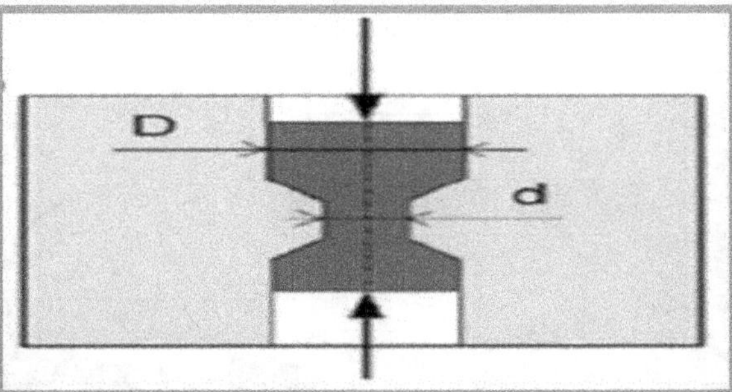

Figure 2.7 Schematic layout of CEC technique [102]

number of deformation cycles. An elevated hydrostatic-pressure is enforced due to cyclic compression and extrusion, resulting in higher loads, which require special tooling. This is the most appropriate process for treatment of soft-material like aluminium alloys.Figure 2.7 Schematic layout of CEC technique (Goebel et al., 2016)Figure 2.7 Schematic layout of CEC technique (Goebel et al., 2016)

2.1.6 Repetitive Corrugation and Straightening Technique

Huang et al. (Sharma & Kumar, 2017) developed the RCS technique for producing the UFG metal/alloys. The schematic layout of RCS technique is represented in Figure 2.7. This process consists of continually bending and straightening of a billet with ridged tooling and then flattening of billet by the application of flat tools without altering the cross- section area of the billet, imposing high-strain into the material. This process depends on the combination of bending and shear that are enforced under constrained-pressure in rolling surfaces which are especially designed. The equivalent-strain per sequence is provided by the following equation:

$$\varepsilon = 4\ln\left(\frac{(t+r)/(0.5t+r)}{\sqrt{3}}\right) \quad \ldots\ldots\ldots\ldots 2.2$$

where, 'r' denotes the curving of bent region and 't' denotes the sample thickness. Repetition of this process in a cyclic manner introduces severe plastic-strain in to the work-piece which results in refinement of grain.

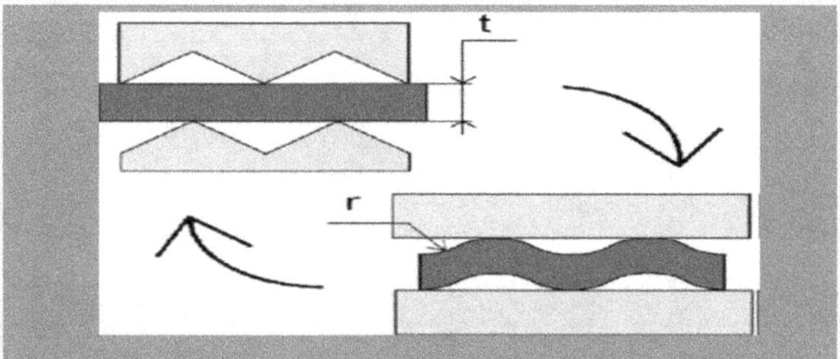

Figure 2.8 Schematic representation of RCS (Sharma et al., 2017)

2.1.7 High Pressure Torsion Technique

Figure 2.8 shows the schematic layout of HPT technique. This process consists the diskette-shaped sample (with diameter range: 10 mm to 20 mm, thinness range: 0.2 mm to 0.5 mm) is compressed in between the anvils with the application of applied pressure (in GPa). Among both anvils one is made stationary while the second one is rotating. Friction developed between the sample surface and anvil causes the monotonically growing shear-strain to be imparted in sample with the rising anvil rotations. The deformation of the sample takes place through compressive stress during shearing and helps in preventing the breaking of sample at very high-strain imposed. In HPT, high pressure can be applied due to suppression of small crack and very satisfactory grain of 100 nm size can be achieved. The significant refinement of structure during HPT process is witnessed after semi or complete (i.e. 360°) rotation of anvil. On the other hand, to achieve the homogeneous nano-structures, deformation by multiple rotations (i.e. complete) are necessary. HPT process can produce homogeneous nanoscale structures including grains-size of about 100 nm or lesser, possessing HAGBs.

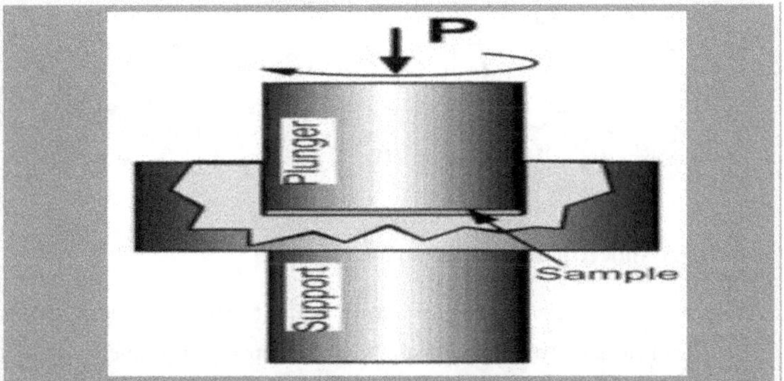

Figure 2.9 Schematic representation of HPT.

2.1.8 Severe Torsion Straining Technique

The STS technique is shown schematically in Figure 2.9 and this process was established by Nakamura et al. (Alam & Sinha, 2019). In STS process, severe torsion strain is induced by rotating one end with respect to another end. The local strain is created by moving the rod along the longitudinal axis of apparatus. Simultaneously, severe torsion strain and local strain causes continuously the huge amount of plastic strain throughout the rod. To create competent STS, the locally heated region should be thin and rod should be rotated faster w.r.t. its moving speed. A modification can be done in cooling-system, which causes heated region to be more limited to a small area to create torsion strain. This technique can be used for amalgamation of powder to produce high density nano- composites and metal-ceramics, and to produce ultrafine-grain and high-strength materials.

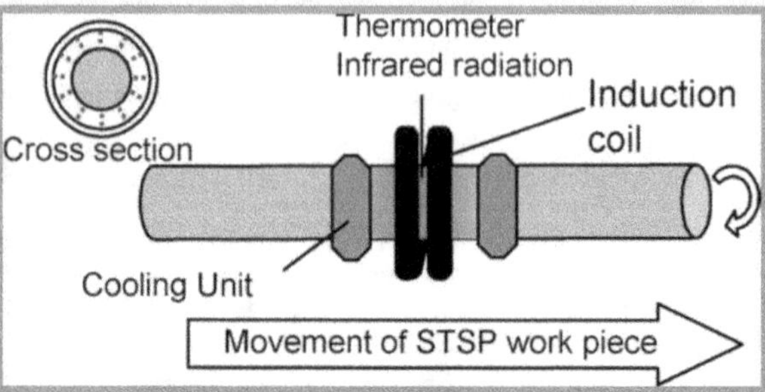

Figure 2.10 Schematic representation of STS (Wang & Zaho, 2020)

2.1.9 Asymmetric Rolling Technique

The schematic representation asymmetric rolling (ASR) process is shown in Figure 2.10 of Kim and Lee (Hajjioui et al., 2022) developed the asymmetric rolling (ASR) process for fabricating the shear texture in the material through shear deformation. Later on, at room temperature the ASR process is applied to produce fine grained structure in the AA 5754 alloy by Jin and Lloyd (Garg et al., 2024) and in pure Al by Cui and Ohori (Gairola et al., 2019). Grain size ranging from 1-2 µm were achieved in AA 5754 due to static recrystallization and in pure Al alloy due to dynamic continuous recrystallization. Differential speed rolling (DSR) is another known form of ASR in which the diameter of the rolls is different or the rotating speeds of the lower and upper rolls are dissimilar from each other. This employs a supplementary shear-strain to the metal/alloy in addition to compressive force (Joshi et al., 2016). It has been successfully used for several metals/alloys for refinement in grains and texture-modifications (Gairola et al., 2019; Hajjioui et al., 2022; Starke, 2014). The advantages of ASR over symmetric rolling are:

i) It can introduce much more redundant shear deformation,

ii) Promotes the deformation to extend to the centre regions of the rolled sheets.

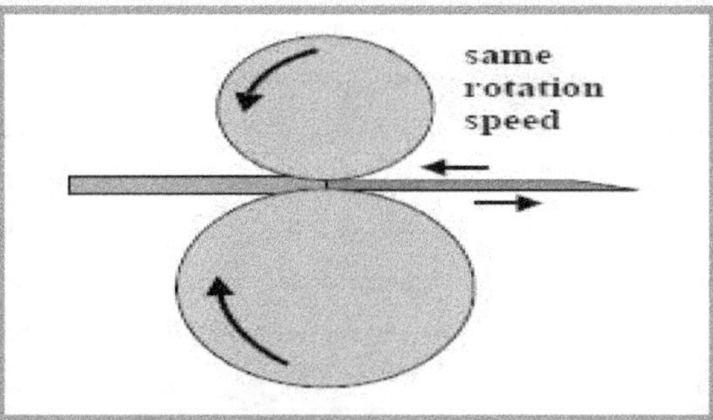

Figure 2.11 Schematic representation of asymmetric rolling (Starke et al., 2014)

2.1.10 Torsion Extrusion Process Technique

Mizunuma et al. (Lynch et al., 2014) developed the principle of TEP technique and schematic representation is shown in Figure 2.11. In this process during an extrusion process the revolution of die/container induces a huge strain in to the metal/alloy. The ductility of metal/alloys raised due to the involvement of high hydrostatic-pressure and this can also introduce the huge torsion-staining in to the workpiece. The equation for getting the average value of strain (representative-strain) on column (cross-section) is as given below.

$$\varepsilon = \frac{4\pi RN}{3\sqrt{3}H} \quad \text{.................. 2.3}$$

Where, 'N' denotes the number of rotations, 'H' denotes column height and 'R' denotes column radius.

2.1.11 Cryo-Rolling Technique

A most proficient technique for producing a UFG microstructure in coarser grained metal/alloys was identified as rolling at liquid nitrogen temperature (LNT or CT: cryogenic temperature) (termed as cryo-rolling process) owing to its various benefits as compared to other SPD techniques. Processing at LNT reduces the amount of strain required for the development UFG microstructure in metals/alloys (Dorin et al., 2018). In addition, at very low temperature, high-density of dislocations/defects can be produced inside the sample due to suppression of DRV resulting noteworthy enhancement in the mechanical characteristics (Babel et al., n.d.). Further, owing to the several advantages such as no requirement of a die,

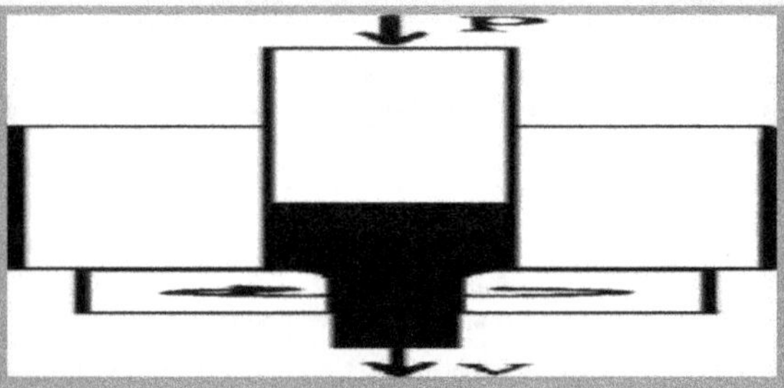

Figure 2.12 Schematic layout of TEP process [108].

Figure 2.12 Schematic layout of TEP process (Lynch et al., 2014).Figure 2.12 Schematic layout of TEP process (Lynch et al., 2014).lesser wear problems, low maintenance cost, simple and economical process associated with cryorolling, makes it highly recommendable process for fabrication of UFG and NC structures in the bulk alloys. Cryorolling can be performed by using conventional rolling machine and liquid nitrogen. The starting material which is to be cryorolled must be in the form of sheet. During the process, samples are dipped into the container of liquid-nitrogen for about 15-30 min. to attain the sample temperature to liquid nitrogen temperature and conventional rolling is performed. The procedures mentioned above can induce huge plastic-strain into the material which leads a significant refinement in microstructure of bulk crystalline solids. Some of them are well established and being used by different research groups to produce bulk

NS or UFG samples of different materials, while others are under development stage and not fully commercialized for large-scale industrial applications due to their limited production rates. It can be observed that all the methods stated above are developed newly within 10 to 15 years. Cryo-rolling, because of its simplicity, less expensive process, no die design requirement, lesser wear problems during processing, low maintenance cost, easy to produce ultrafine and nanocrystalline structure in bulk materials attracts attention of SPD community. Cryo-rolling was first introduced by Wang et.al. (Ji et al., 2021), they deformed copper plate in LNT temperature up to 93% and developed both nanocrystalline and ultrafine grains ranging <300nm with high yield strength. They also reported that annealing at low temperature and at optimized time period, multimodal grain structure with bulk, ultrafine and nanocrystalline grains with large ductility can be obtained without losing strength. A lot of articles are already published on the cryorolling of Cu, Al, Ti, Zr, Steel, Zn, Mg, Ni and their alloys ("What is Aluminum,," n.d.; Panigrahi & Jayaganthan, 2011; Babel et al., n.d.; Ji et al., 2021). represents the schematic representation of various SPD techniques.

2.2 Aluminum Alloys

Aluminium is extensively used in foil and cables. The pure form of aluminium exhibits low weight, high ductility, low yield strength (8-11 MPa only), good corrosion resistance along with good electrical conductivity. The strength of pure aluminium is improved by adding alloying elements to aluminium for different applications. Aluminium alloys are suitable for the low temperature applications. Aluminium forms a very thin layer of oxide over its surface due to its reactive nature with oxygen which acts as an insulator and protects aluminium from external environment. The wrought and cast alloys are the two key categories of Al alloys. The wrought Al alloys are the alloys formed by forging, rolling and extrusion while the cast Al alloys are directly casted into desired shape. Further classification of aluminium alloys falls into two distinct groups, one is categorised as heat treatable Al alloys while the other one is categorised as non-heat treatable Al alloys and are shown in Fig. 2.13.

2.2.1 Heat-Treatable Alloys

Among all Al alloys the mechanical characteristics of few Al alloys can be altered significantly by age hardening and heat treatment process. Al 2xxx, Al 6xxx and Al 7xxx series alloys contains the solute elements in the concentrations that surpass their equilibrium solubility at normal temperature. After solution treatment (dissolution of soluble phase) and water quenching (development of super saturation phase) followed by low temperature ageing, the excess number of solute atoms comes out from the matrix in precipitates form which significantly strengthen the materials. This mechanism of strengthening the material is known as precipitation hardening.

2.2.2 Non-heat Treatable Alloys

These are the alloys which derive their properties from solid solution strengthen and strain hardening. Al 1xxx, Al 3xxx, Al 4xxx and Al 5xxx series alloys are well known under the category of non-heat treatable alloys. The process suitable for the enhancement of mechanical properties is cold working process usually cold rolling and annealing along with thermal treatment required for their stabilization. These alloys are not appropriate for the applications which requires high temperature and high strength because the strengthening properties obtained for the cold worked structures are lost at elevated temperatures.

2.3 Various Strengthening Mechanisms for Al alloys

There are several mechanisms associated with strengthening of Al alloys. Strengthening of aluminium alloys is of great interest because of low density of the alloy, and various people have reported the strengthening through grain refinement of aluminium alloy (Wronski & Bacroix, 2014; Wronski et al., 2013; Kumar, 2015; Lynch et al., 2014). The restriction in the dislocation motion during loading helps in enhancement of strength of the material. There are several ways of restricting the dislocation motion within the microstructure such as precipitates, solute atoms, strain hardening, crystal lattice itself and grain boundaries, thus increasing the density of dislocation in the matrix.

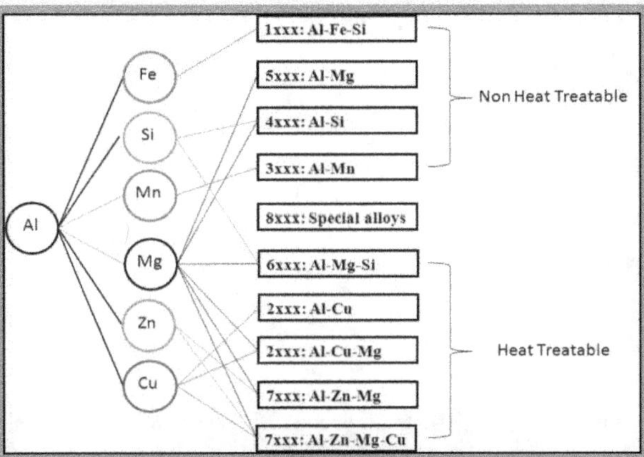

Figure 2.13 Principal alloying elements and the families of alloy derived from them (Rioja, 1998; Ma, Y. et al., 2011)

2.3.1 Strengthening by Grain Size Reduction Submission

In polycrystalline materials, the grain orientation changes abruptly while moving from one grain to another grain across the boundary of grains. The particular slip plane offers difficulty in dislocation movement from one grain to other grain during the deformation. Because of this reason, initially the dislocations during the deformation pile-up near the boundary of grain. The process of piling-up of dislocation is more rapid in smaller grains. In smaller grains the restriction of dislocation motion makes it strain hardened due to which strength increases. If the grain boundary is low angle, grain boundary hindering of dislocations is less, while high angle grain boundary completely hinders the dislocation motion, thus strength of the material increases. The Figure 2.14 (a) shows dislocation moving along a slip plane approaching the boundary of grains.

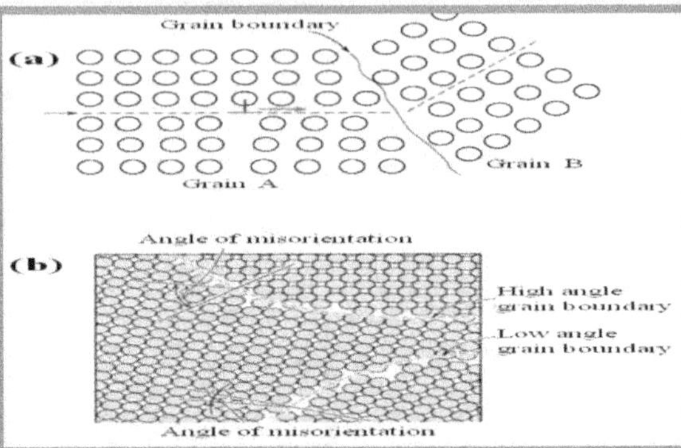

Figure 2.14 (a) Dislocation motion approaching a grain boundary, (b) LAGBs and with respect to adjacent atom positions

The grain size (*d*) influence on yield-strength (σ_y) of material is elucidated by Hall-Petch relation [128].

$$\sigma_y = \sigma_o + kd^{-\frac{1}{2}} \qquad (2.4)$$

Where; 'σy' denotes yield-strength (YS), 'σo' denotes friction-stress, 'k' denotes Hall-Petch constant and 'd' denotes average diameter of grain. From equation (2.4), it is clear that YS of the metal/alloy is inversely proportional to its grain-size; this indicates that reducing grain size helps in enhancing the material strength. The above relation is not valid on very fine-grained and coarse-grained poly-crystalline materials.

2.3.2 Dense-Solution Hardening

'Solid-solution' hardening is a popular strengthening technique for refining the strength of a material, especially for non-heat-treatable alloys. The alloys come under this category are Al 1xxx (Al- small amounts of Si and Fe), Al 3xxx (Al-Mn) and Al 5xxx (Al-Mg) alloys. The applications where a low to medium strength, a good corrosion resistance and a good formability are desirable, these alloys are widely used (Hajjioui et al., 2022). A work on the effect of solid solution on strengthening of the matrix in coarse grained Al alloys was carried out systematically. The addition of stronger atom from another element to the parental atom occupy the substitutional or interstitial positions in the lattice which helps in an improvement of the strength of the parent material. Figure 2.15 (a) and (b) indicates the impact of substitutional and interstitial atoms on the dislocation motion. The factors on which solid solution strengthening depends are shear modulus, size, concentration and valency of solute atoms. The strengthening increment by a solute '$\Delta\sigma ss$' is Figure 2.15 Representation of a dislocation stopped

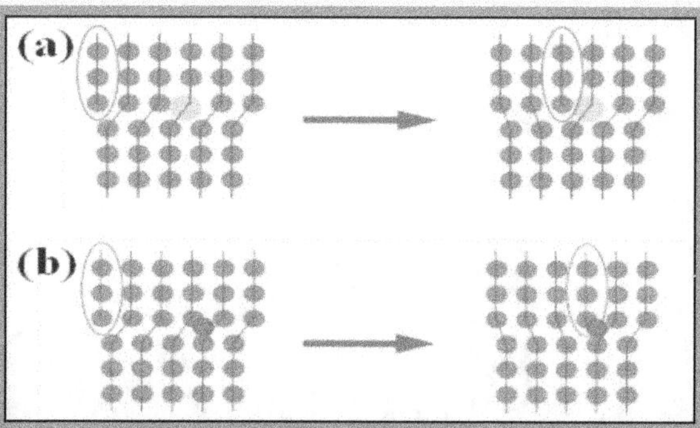

Figure 2.15 Representation of a dislocation stopped by; (a) Substitutional atom (b) Interstitial atom.

by; (a) Substitutional atom (b) Interstitial atom.Figure 2.15 Representation of a dislocation stopped by; (a) Substitutional atom (b) Interstitial atom.defined as;

$$\Delta\sigma_{ss} = HC^{\alpha} \quad \textbf{..........................(2.4)}$$

Where 'C' represents the solute atom concentration, 'H' and 'α' are the constants. A higher concentration of solute atoms will obstruct more dislocation as compared to low solute concentration, thus increasing strength (Wronski & Bacroix, 2014).

2.3.3 Strain Hardening

Strain hardening process is used for improving the strength of the materials with the help of various blacksmiths. Mostly, the wrought alloys are strengthened by the strain hardening technique. The main feature of the plastic-deformation of metal/alloy is the fact that shear-stress which is necessary to yield the slip, rises with enhancing shear-strain [52]. At ambient temperature, the metal/alloy is cold worked and the strain hardening is instigated by faster dislocation interactions, which increases the dislocation density. This increment in the density of dislocation offers obstruction to the movement of dislocation by the crystal lattice. Hardening owing to interaction of dislocation represents the complicated problem due to the involvement of large group of dislocations. The influence of cold-working process on the 'yield stress' and 'ductility' of the metal/alloy is represented Figure 2.16.

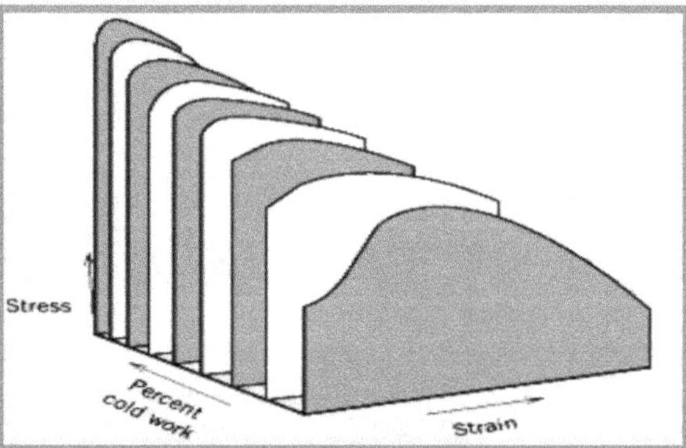

Figure 2.16 The impact of cold-working on yield stress and ductility of a material [127]

2.3.4 Precipitation Hardening Submission

A common source of strengthening of alloy are the tiny second-phase particles (SPP) dispersed in the ductile matrix. Solution-treatment and quenching are the processes used for precipitation hardening. During these processes, at elevated temperature the second-phase is in solid-solution but upon ageing and quenching at lower temperature it precipitates. At elevated temperature, the second phase must be soluble for precipitation hardening to occur but must exhibit diminishing solubility with decrease in temperature (Gairola et al., 2019). In Al alloys of 2xxx series, 6xxx series and 7xxx series, the most useful mechanism for strengthening is precipitation hardening mechanism. The development of fine and uniformly disseminated second-phase particles of solute atoms in matrix causes the enhancement in strength of the Al alloys. The satisfactory smooth precipitates by restricting the gesture of dislocations, strengthen the heat-treatable Al alloys. The three major steps of precipitation hardening process are as follows: (i) Homogenization or Solution treatment of the metal/alloy: In this process, the heating of alloy above the solvus temperature is done along with soaking, until solute dissolves properly and homogenised solid-solution is achieved; (ii) Quenching: In this process the solid solution cools rapidly to a lower temperature producing a supersaturated solid solution and the solute does not immediately diffuse out of a phase; (iii) Finally Precipitation of small size precipitates: In this process the super saturated solid solution heated below the solves temperature. The temperature versus time plot for precipitation hardening mechanism is characterized in Figure 2.17.

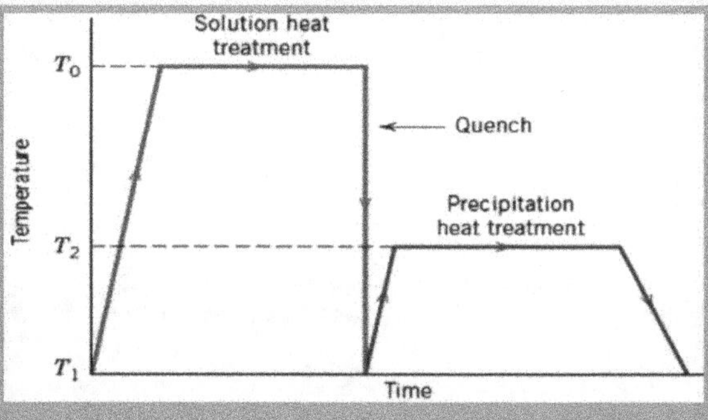

Figure 2.17 Schematic of temperature-time plot showing precipitation hardening Mechanism.

2.3.5 Dispersion Hardening

The small size particles help in cultivating the strength as well as the hardness of metal/compounds by hindering the motion of the dislocation, which escalates the strength of metal/alloy. At advanced temperatures, the SPP can be added in the matrix by simple mixing and consolidation or as precipitates in a solid state. The features like shape, size and amount of SPP helps in controlling the mechanical characteristics like strength of the metal/alloys. The abrupt changes in the dislocation is not able to cut through the precipitate particles even at the high value of stress applied. The moving dislocations bow instead of cutting through the precipitate. The mechanism when the dislocation moves forward by leaving a dislocation loop around each particle is known to be Orowan mechanism (Sevillano, 1980). The schematic representation of Orwan mechanism of dispersion hardening is shown in Figure 2.18.

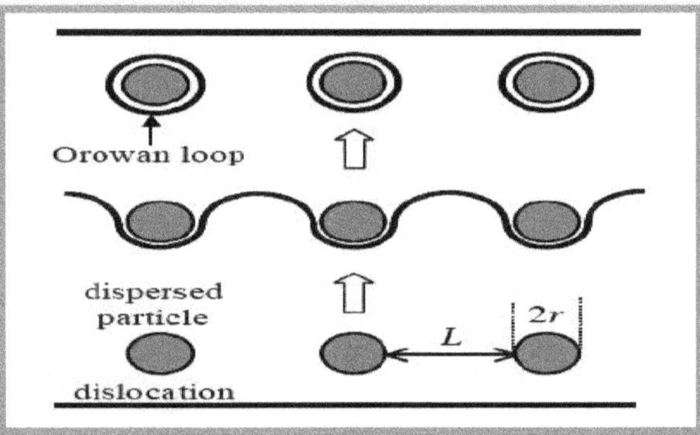

Figure 2.18 Schematic layout of Orwan mechanism for dispersion hardening [133]

2.4 Ductility Enhancement Strategies for UFG Al alloys

Submission Ultrafine-grained (UFG) aluminum alloys formed through 'severe plastic deformation' (SPD) routes exhibit exceptional strength but often suffer from limited ductility due to restricted dislocation activity, rapid strain localization, and limited work-hardening capacity. Enhancing ductility while retaining high strength remains a critical challenge for structural and aerospace applications. This study reviews and evaluates the key strategies developed to improve ductility in UFG Al alloys, including microstructural optimization, processing route design, and alloying approaches. Controlled grain growth through post-deformation annealing, heterogeneous grain structures, and bimodal grain-size distributions have proven effective in promoting strain accommodation and delaying necking. Hybrid SPD techniques such as multi- axial forging combined with cryorolling, 'equal channel angular pressing' with high- temperature passes, and 'accumulative roll bonding' with inter-pass healing enhance dislocation storage and tailor boundary characteristics, resulting in improved uniform elongation. Alloying additions and precipitation engineering—particularly with Mg, Li, Cu, and Sc—strengthen grain boundaries, stabilize UFG structures, and improve strain hardening by precipitate–dislocation interactions. Texture tailoring and grain-boundary relaxation also contribute to improved ductility by enhancing slip system activation. Overall, synergistic integration of heterogeneous microstructures, SPD hybridization, and precipitation strengthening provides a promising pathway to overcome the strength–ductility 'trade-off' in UFG 'aluminum alloys'. This work provides a comprehensive understanding of ductility enhancement mechanisms, offering guidelines for designing next-generation high-performance UFG Al alloys.

2.4.1 Bimodal Grain size distribution

The influence of bimodal-microstructure on ductility without having any effect on strength on nano-structured Cu processed by CR was studied by Wang et al. (Prasad, n.d.) and they experimental that the post-deformation processes such as annealing helps in the improvement of ductility of the 'UFG metals'/alloys. The increased ductility of metal/alloy was attributed owing to the development of micron-sized grains during process of annealing. Subsequently, the numerous attempts were made on various metals/alloys by the researchers to attain the similar type of effects. Nevertheless, the augmentation in ductility was reported with the loss of strength of the metal/alloy. The concurrent enhancement in asset as well as the ductility of

UFG Al 6063 (precipitation hardenable alloy) alloy was reported by Panigrahi et al. (Gairola et al., 2019). They used the following strategy to achieve this enhancement in ductility without the loss of strength: First the ST process followed by the quenching at RT, second the material is deformed using cryo-rolling process to achieve the UFG microstructure, third the short- annealing process for 5 minutes at a temperature of 155 °C and lastly the low- temperature ageing treatment for 12 hours at 125 °C. The deformation of solution- treated metal/alloy by cryo-rolling method results in an enhancement of strength due to the growth of high disruption density as well as the creation of sub-grain boundaries in material. The formation of recovered or partially recrystallised grain structure is observed during the short-annealing process, which acts as a vacant room for newly generated dislocations and helps in accommodating the plastic deformation. Lastly, the ageing at low temperature leads to improvement in strength by the effect of precipitation hardening. The critical parameters to achieve the enhanced ductility without sacrificing the strength during the short annealing process are selection of annealing temperature and time. Although, it is an obvious fact that the short annealing process such restoration through recovery and recrystallization has a clear detrimental effect on the strength of the alloy. In precipitation hardenable alloys, it is not observable due to nullifying effect by cluster strengthening effect. It can be noted that the increase in ductility through this route is with the cost of ductility. Various researchers have reported that the bimodal or inhomogeneous grain structure synthesized by consolidation of mixture of various range of sized powders has shown improved tensile ductility (Valiev et al., 2000; Lavernia et al., 1990).

2.4.2 Introduction of nano-sized precipitates

An effective way to develop the ductility without trailing the strength of UFG Al alloys are the outline of 'nano-sized precipitates into the matrix. The impact of nano-sized precipitates on ductility of 'UFG Al 7075' alloy was investigated by Zhao et al. (Lin & Zheng, 2016) and they identified that the introducing precipitates (nano-sized) helps in improvement of ductility of UFG Al 7075 alloy. They used following strategy for enhancement of ductility in the metal/alloy: (i) Allowing second phase particles to dissolve in to the matrix in coarse grained material; (ii) Producing UFG structure through deformation at supersaturated solid solution state, and (iii) Proper ageing treatments to form precipitates (nano-sized) in the UFG microstructure. Introduction of these precipitates having nano-sized alone will not help in enhancing the ductility of the metal/alloy. The microstructure undergoes the recovery process during the ageing treatment due to which the dislocation density will be reduced. The annihilation of dislocations during recovery process will leave a space for dislocations that generate during deformation. Consequently, increased ductility is attained in metal/alloy. To enable the material to undergo stable plastic flow, the nano-sized precipitates play a major role and also, they accumulate more dislocations; thus, the work-hardening rate of the material gets improved. The drop-in strength due to annihilated dislocations is compensated by the nano sized precipitates, but it also enhances the strength further. Later on, Jiang et al. [144] adopted slightly modified strategy as proposed by Zhao et al., [115] and successfully reported a significant improvement in strength as well as the ductility of AA 2014 alloy. They reported an effective control of precipitation can help in achieving the simultaneous improvement of strength and ductility.

2.4.3 Modification of Grain boundary characters

The microstructure developed during SPD processing can also be characterised by the nature of the grain-boundaries (GBs). The misorientation angle of GBs categorizes it into two categories: if the misorientation with the neighbour grains is less than 15° angle, then it comes under the category of LAGBs and if the misorientation with neighbour grains is greater than 15°, then it comes under the category of HAGBs. In literature, it was stated that the mechanical characteristics of the metals/alloys are directly influenced by the misorientation of the adjacent grains (Ringeval et al., 2006; Aoba et al., 2017; Joshi et al., 2016; Bereczki et al., 2014; Kapil et al., 2019; Kapoor et al., 2013). Grain boundaries render strength to the material by behaving as an obstacle to the gliding of moveable dislocations. Although, the amount of strength that can be imparted to the material is depends upon the grain boundary character. Activation of slip between the adjacent grains is difficult and it requires more energy in the metal/alloy with HAGBs. Thus, the material strength enhances. The rise in strength with the expense of ductility is not desirable as high strength with reasonable ductility of materials is essential for engineering applications. In contrast to HAGBs, LAGBs shows poor resistance to the gliding of mobile dislocations (Joshi et al., 2016). This property can be utilized to improve the ability of plastic deformation in UFG/NC materials. Hu et al., (Ma et al., 2011) have recently reported increased ductility and uniform elongation while retaining high strength by engineering the relative proportion of LAGBs in the Al alloys.

2.5 UFG Material Application

UFG/NC materials can act as the potential materials for structural/functional applications in numerous engineering sectors namely automotive, aerospace, defence, biomedical, constructions etc. (Valiev et al., 2000). The factors which hindering the process of commercialization of UFG/NC metal/alloys are the limiting size of SPD processed metal/alloy and the cost involve in processing (Alam & Sinha, 2019). Although, the recent developments in SPD techniques, allowed scaling up of the processing up to industry production level. 2.5.1 Biomedical applications

The commonly used metals for application in bio-medical field namely body implants are Mg, Fe and Ti. The increasing demand for bio-implants is the main factor for the significant growth of the field of bio-materials. The physical properties that are generally considered for biomaterials includes tensile strength, hardness, modulus, fatigue strength, elongation, impact properties, dimension stability, wear, biocompatibility, biodegradability ("What is Aluminum,," n.d.). Permanent implant materials such as tooth/bone replacements should possess inert behavior in body fluids, whereas for provisional implants, material must degrade at a suitable rate for the aimed applications (Garg et al., 2024). The excellent combination of physical as well as the electrochemical characteristics of Titanium alloys makes them appropriate material for permanent body implants. Implants with pure Fe and Mg are highly biodegradable. Recent studies proved that UFG microstructure obtained through SPD techniques can be used as suitable candidates for biomedical applications (Rout et al., n.d.; MacRae et al., 2018; Goebel et al., 2016). In literature, it is reported that the commercially pure Ti with UFG structure developed through SPD techniques is a better alternative to the Ti alloys. By introducing the nano-sized microstructure in commercially pure Ti, the mechanical characteristics and bio-compatibility have increased (Dhal et al., 2018; Prasad & Malakondaiah, 1992). The materials processed through SPD have shown better corrosion resistance in body fluids over coarse grained materials. Fine grained Mg has shown significant enhancement of fatige life and endurance limit as well as reduction in corrosion rate in Hanks solution (Goebel et al., 2016).

2.5.1 Microelectromechanical systems (MEMS)

In order to have a better mechanical strength of the micro parts in MEMS, the mean size of grain should be less than the part dimension. UFG/NC materials are suitable candidates for these applications. Latest developments in SPD by miniaturizing the processing technique allows to produce micron scale products with desired grain size to increase the both strength and fracture toughness of the metal/alloys (Hajjioui et al., 2022).

CHAPTER 3

LITERATUREREVIEW

This chapter reviews the existing research related to Al–Li alloys and summarizes the experimental studies conducted by various researchers. Different approaches used for the development of ultrafine-grained (UFG) materials are also discussed. In addition, an overview of several severe plastic deformation (SPD) techniques is presented, including their processing methods and synthesis routes. The chapter further examines how these techniques influence the mechanical properties, microstructure, fracture toughness, and anisotropic behavior of materials. In this chapter, a comprehensive overview of the research carried out by various investigators on Al–Li alloys is presented. The chapter discusses experimental studies reported in the literature and summarizes the different approaches used to produce ultrafine-grained (UFG) materials. It also provides a brief discussion of earlier research related to the development and processing of materials using different Severe Plastic Deformation (SPD) techniques. Particular attention is given to how these processing methods influence the overall performance of materials, including their strength characteristics, structural features at the microscopic level, resistance to crack propagation, and the directional variation of properties observed in different materials.

3.1 Aluminium Alloy 2099

AA2099 was selected due to its strong precipitation-hardening potential through phases such as T_1 (Al_2CuLi) and δ' (Al_3Li), along with its good mechanical performance, microstructural stability, and suitability for deformation and cryogenic processing. Thus, it is appropriate for studying microstructural evolution, strengthening behavior, tensile properties, and fracture performance.

S. Ya. Betsofenaet al (Betsofen et al., 2016) proposed a new analytical approach along with suitable equations to estimate the quantity of different precipitate phases formed in aluminum lithium based alloys. Their study focused on the precipitation behavior in 'Al–Cu–Li' and 'Al–Mg–Li' (alloy) systems, particularly the formation of S_1 '(Al_2MgLi), T_1 (Al_2CuLi)', and δ′ (Al_3Li) phases. The findings showed that the amount of δ′ (Al_3Li) strengthening precipitate was much greater in Al–Cu–Li alloys than in 'Al–Mg–Li' alloys. On thc basis of these results, the authors proposed that the strengthening effect commonly assigned to the T_1 (Al_2CuLi) phase in 'Al–Cu–Li' alloys might have been given more importance than it actually deserves. This becomes particularly relevant when the lithium content exceeds nearly 1.5%, as the δ′ phase seems to play a more pronounced role in strengthening.

AlexandreBoisBrochuet al (Bois-Brochu, 2012) The study focused on how extrusion processing affects Al–Cu–Li alloy components, particularly their microstructure, deformation texture, and static mechanical performance. The integrally stiffened panel (ISP) was found to exhibit a predominantly unrecrystallized grain structure, with grains largely aligned along the extrusion direction. In addition, the crystallographic texture was observed to vary across different regions of the panel cross-section depending on the local position. The study also reported the formation of several precipitate phases, including $Al_{20}Mn_3Cu_2$, T_1 (Al_2CuLi), and β′ (Al_3Zr). Based on mechanical testing results, The findings revealed that both yield strength (YS) and ultimate tensile strength (UTS) reach their maximum values along the longitudinal direction, while the lowest strength values were recorded at an orientation of approximately 45°.

D. C. Vander Kooi et al (Vander Kooi, 2000) had experimented and studied mechanical properties of Al-Li alloy C-458 (AA 2099) at cryogenic temperature and its consequences on the tensile and fracture properties on material and found out that there is enhancement in the tensile strength and the ductility at cryogenic temperatures, whereas the fracture toughness of AA 2099 reduced to some extent at 4K.

Nicolas Brodusch et al. (Brodusch et al., 2012) reported that 'field emission scanning electron microscopy' (FE-SEM) is a useful analytical technique for identifying and characterizing the microstructural features of AA2099 and other metallic alloy systems. Their study revealed the presence of several precipitate phases in AA2099, including Al_6CuLi_3, T_1 (Al_2CuLi with Zn and Mg additions), $Al_{20}Cu_2Mn_3$ containing Fe, Al_3Li, Al_2Cu, T_1 (Al_2CuLi), and Al_3Zr with Ti, which were clearly observed through FE-SEM imaging.

D.M. Carrick et al. (Carrick et al., 2013) examined the corrosion performance of AA2099-T8E77 using anodic polarization testing followed by immersion in a 3.5 wt.% NaCl resolution. Their findings presented that this alloy has poorer corrosion resistance than AA2024-T3.In another study,

M. Romios et al. (Romios, 2005) investigated the behaviour of the '2099 Al–Li alloy' subjected to multistage artificial aging combined with natural aging treatments. The findings exhibited that both 'yield strength' (YS) and 'ultimate tensile strength' (UTS) could recover to approximately the same level as the original T861-treated material when a sequential aging process involving two stages and gradually increasing temperature was applied. Additionally, this heat treatment approach resulted in relatively low anisotropy in yield strength.

Y. Maaet al (Ma et al., 2011) The intermetallic phases and precipitate structures in AA2099-T8 5 7 8 1 9 alloy were examined using 'field-emission gun scanning electron microscopy' (FEG SEM), 'transmission electron microscopy' (TEM), and 'electron backscatter diffraction' (EBSD). The microstructural analysis revealed that the elongated grains predominantly contained grain boundaries with large misorientation angles, whereas the nearly equiaxed sub-grain regions were mainly associated with boundaries having smaller misorientation angles. The alloy also contained multiphase particles consisting of both copper-rich and copper-deficient phases, which appeared either separately or in combination. Furthermore, the investigation confirmed the presence of rod-like Al_3Zr dispersoids along with fine age-hardening precipitates such as T_1 (Al_2CuLi), Al_3Li, and Al_2Cu.

Ali Abd El-Atyet al (Abd El-Aty et al., 2018) had reviewed the various strengthening mechanisms and its effect on deformation behaviour, and 'anisotropic' belongings of various 'Al-Li alloys' and found out that there is significant reduction in the anisotropic tensile properties of AL Li alloys along with enhancement in formability, tensile strength, and fracture toughness.

Dipti Samantaray et al (Samantaray, 2010) had done in-depth study on the behaviour of AA 2099 by hot deformation and concluded that it is vital to appropriately design the parameters for thermo-mechanical process as they produce distress in the 'micro structural' evolution and afterward, the 'mechanical properties' of the products.

Xusheng Yang et al (Yang et al., 2017) experimental study the samples deformed under several situations and to investigate their microstructure with their restoration mechanisms. There was noteworthy decrement was observed in the misorientation characteristics and stored energy revealing that the occurrence of dynamic recrystallization (DRX) and these stored energies effects the amounts of dynamic recovery.

Yi Lin et al (Lin & Zheng, 2016) reported the evolution of microstructure of AA 2099 by friction stir welded. Friction stir welding (FSW) give rise to in substantial refining of grains, phases dissolution T1 (Al_2CuLi) along with 'δ′ (Al_3Li), and precipitation' strengthening of 'δ′' phases in the 'nugget zone' (NZ). There was decrement in dislocation density in the welding zones, that give rise to dynamic recovery and recrystallization.

3.2 Grain Refinement and its Effects

M. A. Easton et al (Easton et al., n.d.) had discussed different methods for refinement of grain and several methods for the analysis of grain refinement. His article had developed better understanding of the relationships occurs between grain size, nucleates particle size and their potency along with density of nucleate and the alloy's composition.

Hiromu T animura et al (Tanimura, 1980) studied the traditional Japanese Sword making and investigated its dependency on the distinguishing structure of metallic component of the sword blade i.e. steel. His studies describe the characteristics history related to production of sword blade and their procedure which includes forging followed by quenching and post annealing. In his research effect of grain refinement on various mechanical properties required for perfect Japanese sword had been studied.

R. Z. Valievvet al (Valievv et al., 1986) reported about various phenomenon that are active in the grain boundaries and their interaction with the lattice defects and investigated the importance of grain boundaries in deformation behaviour of metallic compounds and in the solid state physics area i.e. theory of phase transformations in metals, formation of structure and recrystallization.

K. T. Kashyap et al (Kashyap & Chandrashekar, 2001) mentioned about the nucleation and grains growth in Al-Ti alloy and its effects on refining grain and role of solutes grain refinement and observed the formation of α-Aluminium and relationship between solutes present and dendritic growth.

3.3 Multi Axial Forging

K.K. Yogesha et al (Yogesha et al., n.d.) studied the AA 5052's that was processed through multi directional forging (MAF mechanical properties. In his experiments 4.2 cumulative strain was applied and found out enhancement in strength, hardness and ductility along with improved fatigue behaviour. UFG microstructure was been stated as responsible for these enhancements. Presence of nano shear bands was also been seen in his research.

S. Ringevalet al (Ringeval et al., 2006) The study reported a large population of grain boundaries with both small and large misorientation angles in the Al–3Mg–Sc(Zr) alloy after processing by multi-axial forging (MAF). The coexistence of these boundary types contributes significantly to the development of a refined sub-grain structure within the material. The investigation mainly focused on understanding the evolution of crystallographic texture and the changes in microstructural features that occur in the Al–3Mg–Sc(Zr) alloy during the MAF processing route.

Rajat Kapil et al (Bereczki et al., 2014) The effect of 'multi-axial forging' (MAF) on the mechanical performance and fracture toughness of AA8090 alloy was investigated, with a comparison made between samples processed under cryogenic conditions and those forged at room temperature. The results indicated significant grain refinement along with the operation of several strengthening mechanisms, including dislocation strengthening, strengthening resulting from reduced grain size, and hardening associated with precipitate formation, together with a certain contribution from strain hardening. These combined effects contributed to improvements in fracture toughness as well as other mechanical properties of the alloy.

Tomoya Aobaet al (Aoba et al., 2017) observed effects of natural aging o AA 7075 for a long period of time approximately 2.4 years and also artificial aging on overall mechanical properties and development of microstructure via multi-directionally forging (MAF). He concluded the development of 'tensile strength' is due to mechanism of work hardening and artificial aging.He also observed Al18Cr2Mg3and MgZn2 precipitation phases.

Amit Joshi et al (Joshi et al., 2016) A study was conducted on AA2014 alloy to examine the influence of 'multi-axial forging' (MAF) and the role of deformation temperature on its behaviour. The investigation revealed a notable improvement in ductility and hardness along with an growth in the strength of the forged samples, which was mainly attributed to the development of an 'ultrafine-grained' (UFG) structure during processing. The researchers also carried out a detailed comparison of the material's strength characteristics and resistance to crack propagation for specimens processed at room temperature and those forged under cryogenic conditions. The findings revealed that cryogenically forged samples showed better performance than specimens processed at room temperature. It was also observed that fracture toughness improved under both conditions, with a greater increase obtained after cryogenic processing.

P Bereczki1 et al (Kapil et al., 2019) had studied possibilities of grain refinement by the application of cyclic plastic deformation at large-scale on AA 6082 at ambient temperature. Their investigation encompasses of evolution process in mechanical properties by accumulated deformation on the MAFed processed samples. In their observation there is increase in hardness up to a particular no. of passes and then hardness in decreased as number of passes are increased and UFG structure had been observed via TEM.

Kapoor et al. (Kapoor et al., 2013) had reported in his studies the effect aluminium occurs due to the high strain processing. The processing on aluminium is done up to 9 cycles and the softening effect of Al was noticed at that high strain.

3.4 Cryorolling

Cryorolling suppresses dynamic recovery by reducing atomic mobility and limiting dislocation annihilation. This retains high dislocation density, promotes grain refinement and UFG formation, and improves hardness, strength, and overall mechanical performance.

Amit Joshi et al (Yogesha et al., n.d.) The investigation examined AA5052 alloy subjected to several cryogenic rolling routes, including cryogenic rolling, cryogenic groove rolling, and cryogenic groove rolling followed by a subsequent warm rolling stage, with the objective of producing an ultrafine-grained (UFG) structure. The study further evaluated the high-cycle fatigue behaviour of the material after achieving approximately 75% thickness reduction. The specimens processed through cryogenic rolling routes showed better fatigue behaviour than the others. This improvement was mainly due to marked grain refinement, the development of sub-grain structures, and higher dislocation density produced during cryogenic deformation. Moreover, the route involving warm rolling promoted the formation of nanoscale shear bands and fine precipitates, which further improved the mechanical properties as well as the fatigue resistance of the AA5052 alloy.

Nikhil Rangaraju (Rangaraju et al., 2005) The researchers reported a significant enhancement in the overall mechanical performance of cryorolled commercially pure aluminum, including improvements in strength, ductility, and corrosion resistance, which were mainly attributed to the advance of an ultrafine-grained (UFG) microstructure. The study also pointed out that the electrical conductivity of pure aluminum tends to decrease after cryorolling. This reduction occurs because the increase in structural imperfections— such as lattice defects and a larger grain boundary area—creates additional obstacles that interfere with the movement of electrons through the material.

Sushanta Kumar Panigrahiet al (Panigrahi & Jayaganthan, 2011) The authors investigated the behaviour of AA7075 alloy subjected to rolling under cryogenic conditions as well as at room temperature for different levels of strain. The study suggested that cryogenic deformation is an effective method for developing an ultrafine-grained structure, which improves mechanical properties. This improvement was mainly linked to the formation of grain boundaries having high misorientation angles during cryogenic processing. On the other hand, samples rolled at room temperature did not show noticeable recrystallization within the sub-grain structure. The results also showed that cryorolled specimens possessed greater strength and hardness than those processed at room temperature, mainly because of finer grains and the increased dislocation density produced under cryogenic conditions.

Saurabh Gairola et al. (Gairola et al., 2019) examined the outcome of cryogenic rolling followed by annealing on the 'tensile properties' and 'crack growth' resistance of AA8090 alloy. Their observations showed that the cryorolled specimens had higher tensile strength, greater hardness, and better fracture toughness than the solution-treated (ST) samples. However, these gains were accompanied by a reduction in ductility, particularly up to nearly 150 °C. At temperatures above this level, the strength-related properties started to decrease gradually, which was linked to the precipitation of secondary phases such as S′ (Al_2CuMg) and δ (Al_3Li).

Amit Joshi et al. (Joshi et al., 2016) studied AA2014 alloy after cryogenic rolling and subsequent annealing to assess its high-cycle fatigue behaviour. The processed alloy showed an increase in fatigue strength and also exhibited better resistance to crack growth. This improvement was mainly related to lower 'stress concentration' near the crack tip, which helped in improving the fatigue performance of the alloy.

Shokoufeh Malekjani et al. (Malekjani, 2011) reported that processing AA2024 alloy by rolling at cryogenic temperature led to a reduction in low-cycle fatigue strength, while mechanical strength increased and an ultrafine-grained structure was formed.

Manoj Kumar Pathak et al. (Pathak, 2018) examined AA2014 alloy processed through cryogenic rolling followed by warm rolling (CR+WR), with emphasis on its microstructural changes and mechanical behaviour. Their findings showed a clear improvement in hardness, tensile strength, and fracture-related properties. These improvements were mainly associated with the strengthening effect of refined grain boundaries, the simultaneous occurrence of recrystallization and recovery processes, and the influence of dynamic ageing during deformation. In addition, precipitation hardening caused by the formation of fine spherical Al_2Cu particles within the refined grain structure of AA2014 further contributed to the enhancement of the alloy's mechanical behaviour.

3.5 Combined SPD Process

The advantages of combining MAF with cryorolling have been explained. MAF provides multidirectional strain and microstructural homogeneity, while cryorolling suppresses dynamic recovery and retains deformation defects. Together, they promote grain refinement, higher dislocation density, UFG formation, and improved strength, hardness, and stability.

JanuszMajta et al (Majta et al., 2014) reported about the influence of in homogeneity of UFG microstructure progression on SPD specimens prepared by AAD (angular accumulative drawing) and followed by wire drawing and later wire flattening processes. Particularly the study of influence of strain path and micro alloying the elements and deformation in homogeneity on refinement of grain structure and dislocations in the substructure formation processes had been accompanied.

G.H. Salimgareevaet al (Salimgareeva et al., 2006) in depth study on combined SPD process (ECAP+TMT) had been done on Titanium rods. They investigated an innovative and technological method for processing and fabrication of nano structured materials. Severe plastic deformation (SPD) followed by subsequent thermo mechanical treatment, causes the production of bulk billets with high physical and mechanical properties and its influence on micro structural and mechanical properties. The combined SPD process leads to enhancement of the Ti and formation of homogeneous ultrafine-grained (UFG). Moreover, there had been decrease in the anisotropy had been identified. Thus, refinement of grain size result in increment of ductility as well as strength of Ti rod.

K. K. Yogesha et al (Yogesha et al., n.d.) evaluated AA 5052 processed by combined SPD process followed by post-annealing on its fatigue and fracture behaviour. They observed the significant intensification in the HCF life due to the formation of nano shear bands which are also responsible for fragmentation of UFG in AA 5052 processed by combined SPD process followed by rolling. Impurity phase particles are also seen with the presence of fine precipitates formed during warm rolling that are responsible for the growing the HCF strength by effectively dislocation movement. Whereas, fracture toughness had been seen to be enhancing.

Ramesh S et al (Ramesh et al., 2019) had applied combined SPD process i.e. MAF followed by rolling on Cu-Ti alloy at room temperature and its prominence on micro structural evolution, mechanical properties improvement and high corrosion resistance. High shear bands had been analysed in TEM. Whereas, 'low angle grain boundaries' conversion to 'high angle grain boundaries' had been revealed in EBSD. Significant grain refinement with increment in hardness had also been revealed long with high dislocation density.

M. Kulczyk et al (Kulczyk et al., 2007) combines two SPD techniques i.e. hydrostatic extrusion followed by ECAP and applied this technique of combined SPD on Nickel and Copper. Maximum deformation of 13.5 in nickel and 22.3 in copper had applied. They concluded that there had been increment in the homogeneity in the microstructure and thus, improved mechanical properties had been obtained by this process. Therefore, there had been production of nano crystalline materials.

3.6 SPD on AA 2099

A.G. Odeshia et al (Odeshi et al., 2019) examined the mechanical behaviour of AA 2099-T8 when subjected to high strain-rate loading. The response of the alloy was compared with specimens that were cryorolled and subsequently annealed, as well as with ultrafine grained (UFG) samples that were later exposed to multiple impulse momentum. The study revealed the formation of non-equilibrium grain boundaries after the cryorolling process. During dynamic loading, fracture was mainly associated with intense localized shear deformation that produced narrow shear-band regions. In addition to these bands, both deformed and transformed types of shear zones were identified within the microstructure. It was also observed that the high localized temperature generated during deformation caused partial dissolution of precipitates in the transformed shear band region. The findings suggested that precipitation-based strengthening played a major role in growing the yield strength of the alloy, while strengthening due to dislocation activity was also evident.

M.A. Mun˜oz-Morris et al (Muñoz-Morris & Morris, 2010) in their experiment applied ECAP process of SPD on AA 2099 and relate the impact of strain rate with processing temperature, organized with successive annealing treatment on ductility and strength of AA 2099. Strength had been observed to be increased with reasonable amount of ductility under the circumstances where fine precipitates were formed whereas it interrupts the strain localization and fracture in AA 2099.

3.7 High Cycle Fatigue Behaviour

(Amit Joshi, et al., 2016) The present study examined how cryogenic rolling and subsequent annealing influence the 'high-cycle fatigue' performance of bulk 'ultrafine-grained' (UFG) Al 2014 alloy. The alloy sheets were subjected to cryogenic rolling at 'liquid nitrogen' temperature (−196 °C), achieving approximately 75% reduction in thickness. After deformation, the rolled specimens were annealed at temperatures between 100 °C and 250 °C for about 45 minutes to evaluate changes in yield strength, tensile strength, and 10 1 15 fatigue resistance under cyclic loading. Microstructural features of the processed material were studied with the help of transmission electron microscopy (TEM) and field emission scanning electron microscopy (FESEM). The cryogenically rolled material exhibited noticeably better fatigue strength than the solution-treated (ST) condition, mainly because the rolling process produced an ultrafine-grained structure. The refined grains reduce the effective defect size and limit the buildup of stress concentration around crack tips. Furthermore, specimens that underwent cryorolling followed by annealing showed additional improvement in fatigue performance up to an annealing temperature of about 200 °C.

He Fei Li et al (Li et al., 2018) The effect of specimen size on tempered steel's fracture toughness KIC. Their remarks conclude that the size limitation is increased with the improvement in the fracture toughness. Moreover, they analysed the shear fracture mechanisms and flat fracture mechanisms. Their results show establishment of a simple equation that is totally created on the basis of consumed energy through fracture among fracture toughness parameters KIC, Kq, and sample thickness B which reveals a concept that facture toughness (KIC) can be easily attained with small-size specimens. The given relation was stated as substantial application for the evaluation the KICof the materials having high-toughness along with material possessing limited critical size, similar to metallic glass and nano structuredmaterials.

Chao Yin et al (Yin et al., 2019) analysed the transition of ductile to brittle in ITER specification tungsten measured by combination of bending tests and fracture toughness. Their micro structural analysis shows the calculation of the fracture toughness as a function of reasonable accurate temperature.

E.W. Qin et al (Qin et al., 2009) analysed the nano crystalline pure Cu specimens with implanted nano scale twin bundles which were arranged at cryogenic condition by dynamic plastic deformation (DPD). Both increment in fracture toughness as well as tensile strength had been mentioned caused when the DPD parameters were adjusted that led to the increment in the 'volume-fraction' of the nano-twin bundles. The improved strength is accredited to the strengthening mechanism causes due to presence of nano twin bundles. High anisotropic nano-twin lamellar structures were also mentioned. They correlate the improvement in fracture toughness with the development of deep dimples.

E. S. Folias (Folias, 1974) estimated a model approximately for a quasi-elastic plastic analysis and mentioned simple method for the estimation of plastic zone sizes gaining of the crack tip.

Stefan Gollingaet al (Gollinga et al., 2018) reported correlation between the microstructure and fracture toughness of hot stamped boron steel. Their results show that the favourable fracture toughness in bainitic grade whereas brittle mode of fracture had been seen in mixed microstructure bainite and martensite. Also, negligible effect was observed by post heat treatment of bainite.

Ilya A. Ovid'ko et al (Ovid'ko & Sheinerman, 2012) described various mechanisms associated with the improvement in fracture toughness of nano crystalline structured material. They delivered ephemeral summary studies regarding the fracture toughness of nanocrystalline material due to several mechanisms related to plastic deformation. Both conventional deformation mechanisms, i.e. lattice dislocation slip and the noticeable deformation mechanism in nanocrystalline solids i.e. sliding of grain boundary and its migration. It was concluded that formation of GB sliding and its migration led to increase in fracture toughness.

A. Hohenwarter (Hohenwarter & Pippan, 2010) examined the UFG BCC iron's fracture toughness processed via HPT. Fracture toughness is measured along three crack plane orientations. Lowermost value of fracture toughness was measured along with the crack plane into which the inter-granular fracture took place, whereas higher value of fracture toughness has been obtained in other crack plane orientations.

Raviraj Verma et al (Verma et al., 2018) The study investigated the behaviour of Mg–4Zn–4Gd alloy processed through a combination of repeated hot rolling and subsequent hot pressing, commonly referred to as a hybrid severe plastic deformation (HSPD) technique. This processing route was carried out at approximately 450 °C with the objective of producing an ultrafine-grained (UFG) microstructure. The processed material showed higher hardness and better tensile properties. Along with these improvements, fracture toughness was also examined using the J-integral approach, which is commonly applied to evaluate resistance to crack growth. The findings showed that the maximum J-integral values were recorded when the pre-crack was oriented 'perpendicular to the rolling direction' (RD).

3.8 Impact Toughness

(Singh, D. et al., 2015) the study examined the consequence of cryogenic distortion followed by annealing on the microstructure and mechanical properties of Al-5083 alloy. Rolling under liquid nitrogen caused a clear increase in both hardness and strength. The ultimate tensile strength was about 340 MPa for the sample with 30% thickness reduction and nearly 390 MPa for the sample with 50% reduction. These rolled specimens were later subjected to Charpy impact testing in the (temperature range) from −190 °C to 100 °C. The results showed that increasing the amount of thickness reduction during cryorolling raised the yield strength but reduced ductility, which consequently lowered the impact toughness at all testing temperatures. Annealing after rolling significantly improved the impact toughness because of recovery and recrystallization. After annealing at 350 °C for 1 hour, the (average grain size) of the 50% reduced specimen was around 14 μm, whereas the

30% reduced specimen showed a larger grain size of nearly 25 μm. Fractography by scanning electron microscopy showed large dimples in the untreated alloy, indicating ductile fracture, while the cryorolled samples exhibited a fibrous fracture surface with much finer dimples, suggesting a shift toward relatively brittle fracture.

Mohammad Tajally et al., accepted a comparative learning on the effect of cold deformation and subsequent 'heat treatment' on the 'tensile properties' and 'impact toughness' of 7075 aluminium alloy. The cold-rolled samples were annealed at (temperatures ranging) from 225 °C to 345 °C for 5 minutes. A marked improvement in ductility and impact toughness was observed when the annealing temperature rose above nearly 265 °C. The study also showed that plastic deformation by cold rolling had a strong influence on the directional variation of strength, and this anisotropy became more evident with increasing cold work. The maximum strength was recorded in the transverse direction. Although cold rolling improved the strength of the alloy, it reduced the impact toughness, and the reduction became larger as the deformation level increased. Even after recrystallization during annealing, the absorbed impact energy could not be fully recovered. The authors further pointed out that impact testing can be used as a simple method for evaluating toughness and plastic anisotropy in sheet and plate materials. Fracture surface examination through scanning electron microscopy showed large dimples in the undeformed alloy, confirming ductile fracture, whereas the cold-rolled specimens displayed a fibrous surface along with quasi-cleavage regions, indicating a transition from ductile to relatively brittle fracture.

3.9 Low-Cycle Fatigue (LCF) Behaviour

Atasi Ghosh explained that the 'low-cycle fatigue' behaviour of engineering metals is strongly controlled by the 'microstructural' changes that take place during repeated plastic straining. Under cyclic loading, the internal structure of a material gradually changes, and these changes directly affect its mechanical response. The review describes the basic mechanisms involved in cyclic deformation in different metallic materials. It particularly discusses cyclic hardening and softening, the development and interaction of dislocations, and the role of crystal structure and anisotropy in FCC and HCP metals. The review also compares a number of theoretical and computational approaches that are commonly used to represent plastic deformation and fatigue behaviour during cyclic loading. Contreras investigated the 'low-cycle fatigue' behaviour of an ultrafine-grained AA5083 'aluminium alloy' reinforced with boron carbide particles and produced through a cryomilling route. The processed alloy developed a trimodal microstructure made up of coarse grains, ultrafine grains, and uniformly distributed reinforcing particles. This mixed structure had a strong effect on the cyclic deformation behaviour, including both hardening and softening during repeated loading. The consequences showed that the 'fatigue strength' of the alloy with trimodal microstructure was similar to, and at some strain amplitudes even better than, that of the conventional coarse-grained alloy.

Paul J. Vardanega reported that recent work has focused on predicting the 'low cycle fatigue' life of structural aluminium alloys by using advanced modelling techniques. One such study dealt with 6000-series aluminium alloys tested at large strain amplitudes of about ±4 to 6%. In that investigation, a calibrated constitutive model combined with a fatigue damage formulation was used, and the predicted results were in close agreement with the experimental findings. This showed that the approach can be useful for the design and safety assessment of aluminium structural members subjected to repeated loading, especially in seismic applications. In another study, Jagtar Singh (2022) examined the effect of prior tensile deformation and different strain ratios on the low-cycle fatigue response of alloys such as Al 6061 and Al 2024. The study showed that pre-deformation generally reduced fatigue life. It was also observed that the cyclic stress-strain response involved both hardening and softening, depending on the strain amplitude and mean strain applied. These results are important for components that experience plastic deformation during manufacturing before actual service use.

Jacques Stolarz studied the 'low-cycle fatigue' behaviour of two-phase aluminium-silicon alloys to gain insight into the development of damage during cyclic loading. These alloys contain a ductile aluminium matrix along with relatively brittle silicon particles, and this combination has a major influence on fatigue performance. It was observed that crack initiation often starts near the brittle silicon particles because they cannot accommodate plastic deformation effectively. As a result, the fatigue life of the alloy depends not only on the overall cyclic deformation response but also on the local mechanisms governing crack initiation and crack growth. These outcomes demonstrate the importance of microstructural heterogeneity in controlling the fatigue resistance of Al-Si alloys.

CHAPTER 4
OBJECTIVES

The thesis has been undertaken with the following objectives:

- Formation of UFG structure of Aluminium-Copper-Lithium alloy 2099 through combined SPD processes i.e. processing of material under any SPD process followed by other SPD process. In present thesis AA 2099 is processed under Multi Axial Forging followed by Cryorolling.

- Studying the effects of combined SPD processes on micro structural and mechanical properties of AA 2099.

- Evaluation of the outcome of post-annealing on the 'mechanical properties' and fracture characteristics of AA2099.

- To examine the influence of multi-axial forging and cryorolling, followed by post-annealing treatment, on the 'high-cycle fatigue' behaviour and impact toughness of AA2099 alloy.

- To correlate the mechanical properties and micro structural features by Optical microscopy, Vickers Hardness, UTM, High Cycle Fatigue SN Curve, charpy, izod test and SEM studies.

- To select the processing condition that results in the most favorable combination of mechanical behaviour, high-cycle fatigue resistance, and impact toughness for AA2099 alloy.

CHAPTER 5

EXPERIMENTAL PROCEDURE

5.1 Procedure Adopted for Experimentation

5.1.1 Processing

In the present work, AA2099 alloy plates were procured from a Mumbai-based supplier in India. The alloy was provided in the form of sheets with approximate dimensions of (255 × 355 × 45) mm

5.1.2 Solution Treatment of Procured AA 2099

Initially, specimens with dimensions of approximately (20 × 25 × 40) mm. were machined from the AA2099 alloy sheet. To obtain an ultrafine-grained (UFG) structure, the specimens were processed through multi-axial forging (MAF) followed by cryorolling (CR) under different temperature conditions. Several material states were examined in this study, including the solution-treated condition, specimens processed by MAF followed by CR, and samples that were additionally subjected to post annealing treatment at temperatures of 150 °C and

Figure 5.1 Electric muffle furnace heating ST samples to 540°C

200 °C after the deformation processes. Prior to these thermomechanical treatments, all specimens were exposed to a solution treatment temperature of 540 °C for two hours and then rapidly cooled rapidly using water as the quenching medium at room temperature in order to obtain a uniform and 'equiaxed microstructure'.

5.1.3 Combined SPD (MAF+CR) Process

The prepared specimens were subjected to multi-axial forging (MAF) for a total of six cycles. After each deformation cycle, the specimen were cooled by immersion

Figure 5.2 Forging friction screw press and Rolling machine for multi axial forging (MAF) and C

in cryo nitrogen at approximately −190 °C for about 15 minutes to maintain cryogenic conditions. During the MAF operation, deformation was carried out using an open-die forging setup. In each pass, the specimen was compressed along its longest dimension and then reoriented by 90° before the next pressing step; this sequence constitutes a single pass. A true strain of about 0.2 was introduced in each pass, resulting in an accumulated strain of approximately 0.6 after compression along the three orthogonal directions, which is considered one complete cycle. The same procedure was repeated to obtain higher levels of accumulated strain, reaching approximately 1.2 after two cycles, 2.4 after four cycles, and 3.6 after completing six cycles. Subsequently, specimens measuring approximately 20 mm × 25 mm × 10 mm were prepared from the central region of the MAF-processed material along the z-direction of the original 40 mm thick block. These specimens were then subjected to cryogenic rolling. Before each rolling pass, the specimens were immersed in liquid nitrogen for nearly 15 minutes in order to maintain cryogenic conditions during deformation. Through this rolling process, a thickness reduction of about 90% was obtained, corresponding to a true strain of approximately 2.3. Similar experimental procedures have been adopted by several researchers to determine the influence of cryogenic rolling and subsequent heat treatment on the mechanical response of different alloys (Valievv et al., 1986; Kashyap & Chandrashekar, 2001; Mohanty & Gruzleski, 1995). After completion of the cryorolling process, the plate-shaped specimen were thermally processed at temperatures of 150 °C and 200 °C for a duration of one hour and were then rapidly cooled in water at room temperature.

5.2 Sample Preparation

5.2.1 Optical Microscopy Samples Preparation Optical microscopy of ST and deformed samples was performed by Leica DMI 5000 under the polarized light with the help of poultens reagents. Further, the next step in the preparation for OM samples is followed by polishing the surface for (ST, MAF+CR, MAF-CR) sample heated to 150°C and (MAF-CR) sample heated to 200°C.Polishing of the specimens were carried out through emery papers of multiple grades used in a sequence of 320, 800, 1200 and 1500. Once this polishing gets completed by these emery papers

Figure 5.3 Optical Samples in series of three samples each for ST (1), MAF+CR (2-3), MAF+CR+PA+150°C (4-5) and MAF+CR+PA+200°C (6-7) respectively

as mentioned the samples were passed through the next step of cloth polishing. At the time of cloth polishing MgO (Magnesium oxide).

5.2.2 Preparation of Mechanical Testing Samples

The mechanical behaviour of the Al 2099 alloy was evaluated by conducting tensile and hardness measurements. Tensile testing was carried out using an H25K-S Tinius Olsen universal testing machine, where the test was conducted by loading the specimen at a fixed crosshead speed corresponding to an initial strain rate of approximately 5×10 (s . The preparation and dimensions of the tensile specimens followed the specifications provided in the ASTM E8 standard Sub-size tensile specimens were fabricated with a gauge length of approximately 10 mm. The dimensional details and geometry of the tensile specimens used in the present investigation are provided in the following specifications. shown in Figure 5.3.

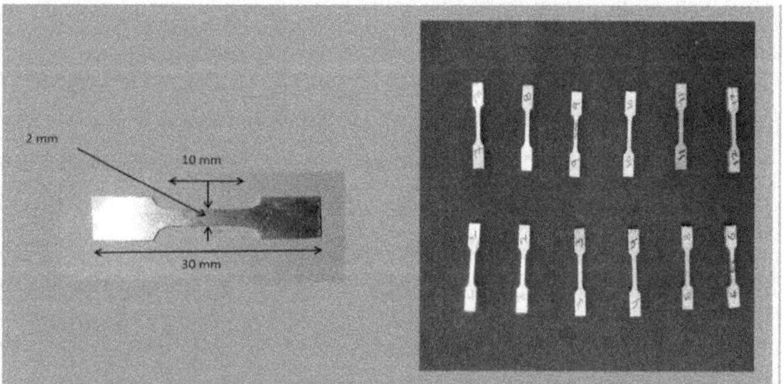

Figure 5.4 Tensile Samples in series of three samples each for ST (1-3), MAF+CR (4-6), MAF+CR+150°C (7-9) and MAF+CR+200°C (10-12) respectively

5.2.3 Scanning Electron Microscope (SEM)

Fractured surface obtained after high cycle fatigue test impact toughness test had been analysed through FESEM.

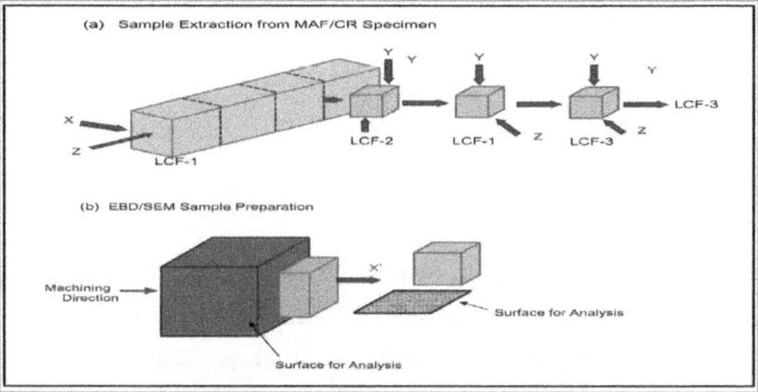

Figure 5.5 Scanning Electron Microscope Sample Abstract Process (SEM)

5.2.4 Hardness Test

The hardness measurement of the was carried out AA2099 alloy was measured using a Vickers hardness tester with an applied under a 5 kg load and a dwell time of 15 seconds. For each specimen, three indentations were made at different positions on the surface to ensure reliable results. The hardness value reported for each sample represents the average of these measurements.

5.2.5 Fatigue Test In The 'High-Cycle' Regime

'High-cycle fatigue testing' of the specimens was conducted with the help of an 4 Instron universal testing machine. The samples prepared for fatigue experiments were fabricated from the processed material with their longitudinal axis aligned in alignment with the rolling direction, following the specimen geometry specified in ASTM E466-07. After machining, the gauge sections of the samples were carefully polished with emery papers up to 1500 grit in order to eliminate minor surface imperfections introduced during machining. Fatigue loading was applied in the axial direction under stress-controlled conditions, while keeping All specimens were subjected to fatigue loading at a stress ratio of 0.1 and a constant frequency of 20 Hz.

5.2.6 Impact Toughness Test

Impact resistance was assessed using Izod and Charpy tests on all prepared specimens to evaluate their impact toughness. For the Charpy test, standard specimens were prepared in accordance with the specified testing guidelines. Each specimen consisted of a rectangular bar with dimensions of the specimens were fabricated with approximate dimensions of 55 × 10 × 10 mm, and a notch was machined on one of the larger faces to promote crack initiation. initiation during impact loading. A notch was introduced on one of the wider faces : 2mm deep, with

Figure 5.6 Impact energy testing samples for ST (1-3), MAF+CR (4-6), MAF+CR+PA+150°C (7-9) and MAF+CR+PA+200°C (10-12) respectively Where S = 55mm, W = 10 mm, a = 10 mm and notch: 2mm deep, with 45° angle and 0.25mm radius along the bas

45° angle and 0.25mm radius along the base

5.2.7 Low-cycle fatigue experiments:

Low-cycle fatigue specimens for all sample conditions the test specimens were extracted from the deformed material along the rolling direction. in accordance with ASTM standards E606 (strain-controlled LCF practice) Use a cylindrical, reduced-gauge specimen to get a uniform axial strain field in the gauge length. Gauge diameter, d = 6 mm, Gauge length, Lo = 2 d = 12 mm Reduced (parallel) section length (Lr): keep ≤ 8 d (i.e., ≤ 48 mm). Transition radius, r = 2 d = 12 mm Grip (end) diameter, D = 2 d = 12 mm. Al-Li alloys are temper-sensitive; always test the actual condition. Remove rough-turn to near net shape Final surface finish target: Ra ≤ 0.8 μm After turning, remove a thin layer (0.05 0.10 mm) by light polishing or fine turning to eliminate any plastically-deformed surface layer introduced by grinding. Al-Li alloys are prone to surface damage, so this step is important. Polish gauge with progressively finer abrasive (e.g., 320 → 600 SiC paper if needed), then cloth polish with 1 μm diamond / 0.25 μm diamond or alumina test parameters Al-Cu-Li-2099 total strain control (strain-controlled LCF, per ASTM E606 Strain amplitudes: choose a set covering the LCF range observed for Al-2099 Typical study points: 0.3%, 0.4%, 0.5%, 0.6%, 0.7%. Fit Coffin-Manson/Basquin to results. Frequency: 0.5–1 Hz specimen fracture Document clearly. 6 specimens per strain level Post-test handling & metallography After failure, carefully capture fracture surfaces for SEM (clean gently; do not polish fracture face). Cut cross-sections

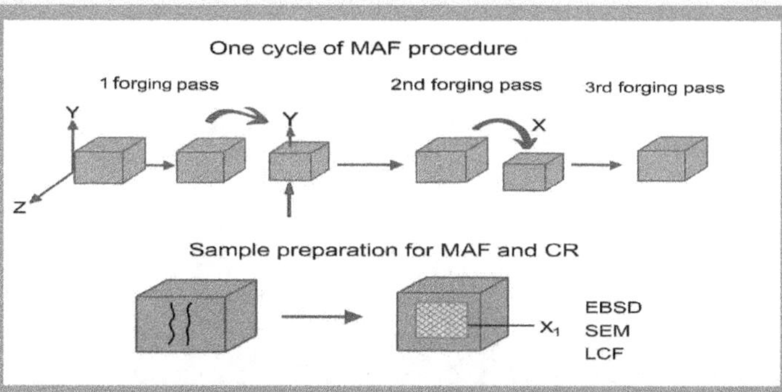

Figure 5.7 sample for low cycle

across gauge near initiation site for metallography (mount, grind/polish, etch appropriate to reveal microstructure). Correlate local microstructure (precipitates, grain boundary features) with failure mode. Al-Li alloys often show intergranular features and precipitate influence on crack initiation — document temper & local hardness.

CHAPTER 6

RESULT AND DISCUSSION

This chapter is chiefly concerned with includes the outcomes of various microstructural characterization and mechanical properties obtained after combined SPD MAF+ The specimens were subjected to heat treatment at two different temperatures. 150°C and 200°C for a period of 45 minutes to examine microstructure. This chapter is about the detailed discussion of the various testing methods and result obtained by these testing.

6.1. Optical Microstructure

The optical microstructure of the Al 2099 is presented in Figure 6.1. solution-treated AA2099 alloy. The micrograph reveals a fairly uniform distribution of 'equiaxed grains' with the 'average grain size' ranging from 100 to 130 μm. Figures 6.1(b) to 6.1(d) show the microstructures of the samples processed through the hybrid SPD route and those subsequently annealed at 150 °C and 200 °C. In these images, the microstructure appears highly distorted, with elongated and fragmented features aligned along the rolling direction. Figure 6.1(b), which represents the MAF+CR condition, exhibits a heavily deformed structure. Because of the intense fragmentation, the grain boundaries are not clearly distinguishable from the grain interiors. For the same reason, accurate grain-size measurement is not possible from the optical micrograph alone.

After 'post-annealing' at (150 °C and 200 °C,) the alloy shows a gradual recovery of the deformed structure, as seen in Figures 6.1(c) and 6.1(d). Although the distinction between grain interiors and boundaries remains unclear because of the refined nature of the microstructure, the extent of recovery becomes more noticeable after annealing. In the MAF+CR sample, only a limited number of recovered grains can be identified. With annealing, however, the fraction of recovered grains increases, and this effect becomes more evident as the annealing temperature rises from 150 °C to 200 °C. This behavior suggests that recovery processes become more active at higher annealing temperatures, leading to a gradual reduction in dislocation density. Since the optical micrographs do not provide sufficient detail to fully reveal the refined structure, TEM analysis was also carried out and is discussed in a later section.

The microstructural evolution of the solution-treated AA2099 alloy, as well as that of the alloy subjected to the hybrid SPD route (MAF+CR) and subsequent annealing at 150 °C and 200 °C, was examined by TEM and is presented in the figure. 6.1

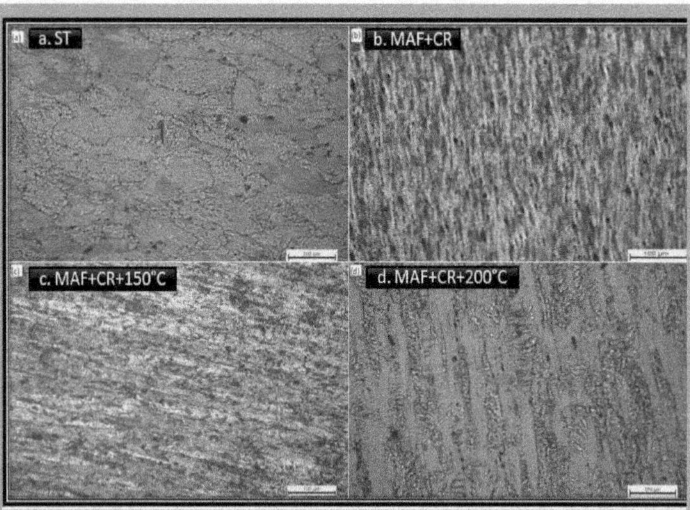

Figure 6.1 Optical Microstructure of Al 2099 (a)ST (b) MAF+CR (c) MAF+CR+PA at 150°C (d) MAF+CR+PA at 200°C

6.1.1 Microstructural characterization was carried out using 'Transmission Electron Microscopy' (TEM).

The microstructure of the solution-treated AA2099 alloy underwent noticeable changes and, in the alloy, processed through the hybrid SPD route, namely MAF+CR, MAF+CR followed by post-annealing at 150 °C, and MAF+CR followed by post-anncaling at 200 °C, were examined using TEM. The corresponding TEM micrographs for the ST, MAF+CR, MAF+CR+PA at 150 °C, and MAF+CR+PA at 200 °C conditions are presented in Figure 6.2(a)–(d), respectively. Figure 6.2(a) represents the TEM image of the solution-treated alloy. In this condition, the microstructure appears relatively clean and does not show any significant deformation-related defects. After the application of the hybrid SPD process, the alloy undergoes intense plastic deformation. Multiaxial forging introduces a high level of strain and generates a large number of dislocations, while the cryorolling step suppresses dynamic recovery. As a result, the dislocation density increases markedly, particularly along low-angle grain boundaries. In Figure 6.2(b), dense dislocation tangles and fine dislocation cell structures can be clearly seen. The micrograph also indicates that the material is in a heavily deformed state with a high concentration of dislocations distributed throughout the

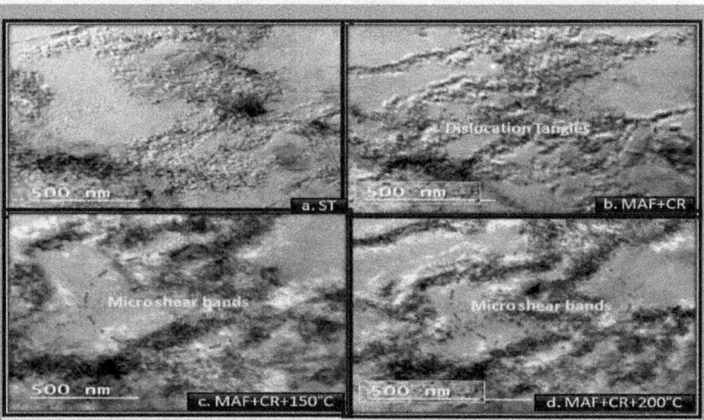

Figure 6.2 TEM Microstructureof Al 2099 (a)ST (b) MAF+CR (c) MAF+CR+PA at 150°C (d) MAF+CR+PA at 200°C

matrix. When the hybrid SPD-processed samples are subjected to post-annealing at 150 °C and 200 °C, gradual recovery of the deformed structure begins to occur. In these conditions, the subgrains formed within the dislocation-rich regions start to rearrange and develop into more defined boundaries. Because of this process, micro-shear bands or angular boundary features begin to appear in the micrographs, as shown in Figure 6.2(c) and Figure 6.2(d).

6.1.2 The grain structure and boundary characteristics were examined by Electron Backscatter Diffraction (EBSD).

EBSD was carried out to examine the changes in microstructure. in solution-treated AA2099 alloy and in the material processed under different conditions, namely MAF+CR, MAF+CR+PA at 150 °C, and MAF+CR+PA at 200 °C. The corresponding EBSD maps are presented in Figure 6.3(a)–(d). The solution-treated sample in Figure 6.3(a) reveals an elongated grain structure with an average grain size of approximately 130 μm. Figures 6.3(b)–(d) show the EBSD micrographs for the MAF+CR condition and the two annealed conditions. In Figure 6.3(b),

Figure 6.3 Electron backscatter diffraction (EBSD) (a) ST (b) MAF+CR (c) MAF+CR+PA at 150°C (d) MAF+CR+PA at 200°C

elongated cell bands aligned with the rolling direction are clearly visible, and the boundary misorientation is mainly distributed within the range of 2° to 15°. The microstructure in this condition consists of long fibrous grains parallel to the rolling direction, finer grains, and a relatively large fraction of low-angle grain boundaries. The microstructures after post-annealing at 150 °C and 200 °C are shown in Figures 6.3(c) and 6.3(d), respectively.With annealing, part of the dislocation-rich elongated structure begins to rearrange, leading to the development of subgrains in some regions. At the same time, many of the elongated grains become further fragmented, and their thickness is reduced to the submicron range. In comparison with the solution-treated and MAF+CR conditions, the post-annealed samples do not show a large population of fully equiaxed grains; instead, they still contain a considerable fraction of elongated subgrains with relatively low aspect ratio. The EBSD maps also indicate that the elongated grains in the annealed samples are shorter than those in the other conditions. 1 Based on the EBSD observations, the MAF+CR+PA at 150 °C condition appears to produce the most pronounced grain refinement among the processed samples, reducing the structural scale to the submicron level.

6.1.3 Boundary Misorientation histogram

Figures 6.4(a)–(d) present the histograms showing the subgrain misorientation the distribution for the solution-treated, MAF+CR, MAF+CR+PA at 150 °C, and MAF+CR+PA at 200 °C sample conditions. The solution-treated sample shown in Figure 6.4(a) contains a comparatively larger fraction of high-angle grain boundaries (HAGBs). In contrast, the MAF+CR condition shows a predominance of low-angle grain boundaries (LAGBs), indicating the strong effect of deformation on the microstructure. With the application of post-annealing, the boundary character begins to change. As the annealing temperature increases, the fraction of HAGBs rises, while the proportion of LAGBs gradually decreases. This trend suggests the occurrence of recovery and recrystallization processes during annealing. In aluminum alloys, such changes are generally associated with the high deposited energy introduced by 'severe deformation' particularly in the form of a

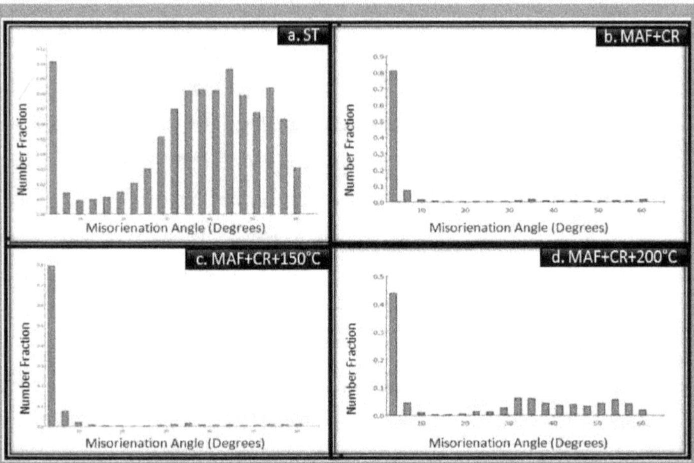

Figure 6.4 Histograms (a) ST (b) MAF+CR (c) MAF+CR+PA at 150°C (d) MAF+CR+PA at 200°C

high dislocation density, which provides the driving force for microstructural rearrangement. A comparison of all four histograms shows that the MAF+CR sample contains the highest fraction of LAGBs. After annealing at 150 °C, this fraction decreases slightly, and a further shift is seen after annealing at 200 °C, where the increase in HAGBs becomes more pronounced. These observations are consistent with earlier reports by Panigrahi et al. for Al 7075 alloy processed by room-temperature and cryogenic rolling.

6.2 Mechanical Properties

The main focus of this includes the outcomes of various mechanical properties obtained after combined SPD MAF+CR samples were heat treated for the at two annealing temperatures, 150 °C and 200 °C, for a duration of 45 minutes to evaluate the results.

6.2.1 Vickers Hardness

Figure 6.5 presents the Vickers hardness values of AA2099 alloy under the hardness of the material was evaluated under the ST, MAF+CR, MAF+CR+PA at 150 °C, and MAF+CR+PA at 200 °C conditions. For each condition, five specimens were tested to ensure reliable measurement of hardness. The solution-treated alloy showed a hardness of 108 HV (±1.6). After hybrid SPD processing through MAF+CR, the hardness increased to 154 HV (±0.66), representing an improvement of about 43% over the ST condition. A further increase in hardness was observed after post-annealing at 150 °C, where the maximum value of 184 HV (±1.1) was obtained. When the annealing temperature was raised to 200 °C, the hardness

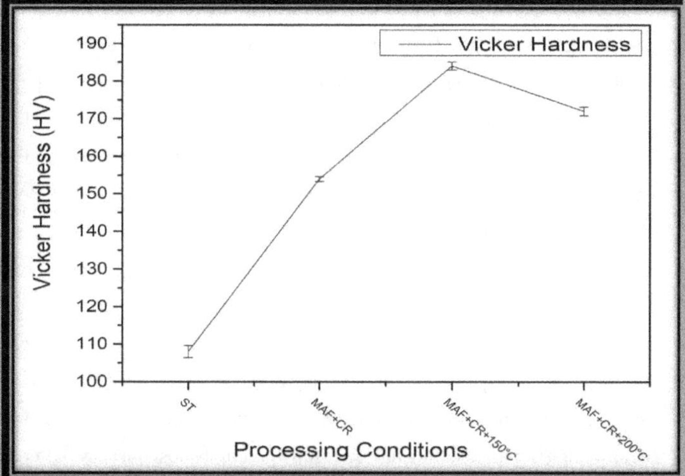

Figure 6.5 Variation in Vickers Hardness at different processing conditions

decreased slightly to 172 HV (±1.27), although it remained higher than that of the ST and MAF+CR conditions. 7 1 12 1 The initial rise and subsequent reduction in hardness after annealing can be explained by the combined influence of several 'strengthening mechanisms' including grain 1- boundary strengthening, 2- strain hardening, dislocation strengthening, and 3- precipitation hardening. The higher hardness after MAF+CR and annealing at 150 °C is likely related to the formation of an ultrafine-grained structure and the restricted recovery of the deformed microstructure. At 200 °C, partial recovery and microstructural relaxation may reduce the strengthening contribution, leading to a slight drop in hardness.

6.2.2 Tensile Strength

Figure 6.6 and Table 6.1 present the tensile properties of AA2099 alloy in the ST, MAF+CR, MAF+CR+PA at 150 °C, and MAF+CR+PA at 200 °C conditions. The solution-treated alloy showed a yield strength of 196 MPa and an ultimate tensile strength of 402 MPa. After processing through the MAF+CR route, both strength values increased markedly. In this condition, the yield strength reached 412 MPa, while the ultimate tensile strength also increased significantly. The highest tensile strength was obtained for the MAF+CR+PA at 150 °C condition. This sample exhibited a yield strength of 475 MPa and an ultimate tensile strength of 547 MPa. When the post-annealing temperature was increased to 200 °C, both values decreased to some extent, with the yield strength falling to 436 MPa and the ultimate tensile strength to 481 MPa. Even so, these values remained higher than those of the ST condition. In terms of ductility, the solution-treated sample showed the largest elongation, with a value of 18%. After hybrid SPD processing, the alloy experienced a sharp drop in elongation, which decreased to 3.9%. This reduction reflects the strong strain hardening and structural refinement produced by severe deformation. Post-annealing helped restore part of the lost ductility. After annealing at 150 °C, the elongation increased to 8.3%, and a further rise to 12.3% was observed after annealing at 200 °C. Overall, the results indicate that the MAF+CR+PA at 150 °C condition offers the most favorable combination of tensile properties, as it provides the highest strength while retaining a moderate level of ductility. Various researches (Yogesha et al., n.d.; Joshi et al., 2016; Pathak, 2018; Yogesha et al., n.d.; Odeshi et al., 2019) had reported the first decrement in the ductility of material after SPD process and further improvement in their ductility as the showing below figure.

Table 6.1 Tensile Strength data

Processing Condition	YS (MPa)	UTS (MPa)	% Elongation	Hardness (HV)
ST	196	402	18	108
MAF+CR	412	501	3.9	154
MAF+CR+150°C	475	547	8.3	184
MAF+CR+200°C	436	481	12.3	172

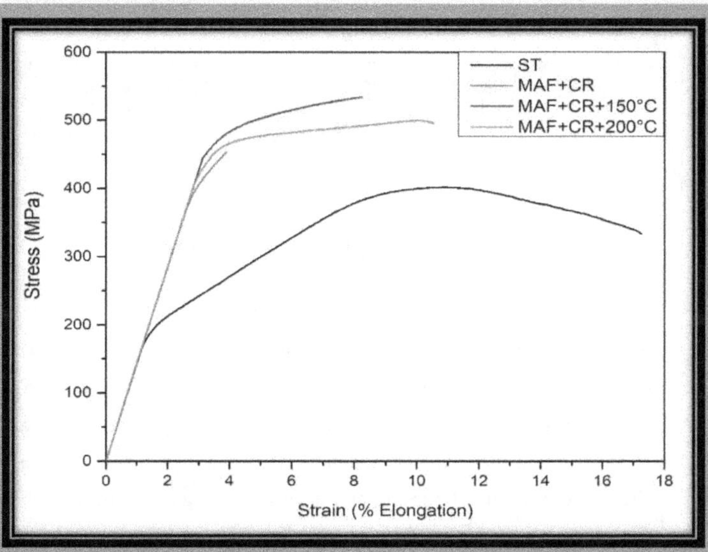

Figure 6.6 Variation in Stress-Strain Relationship in AA 2099 after various processing conditions

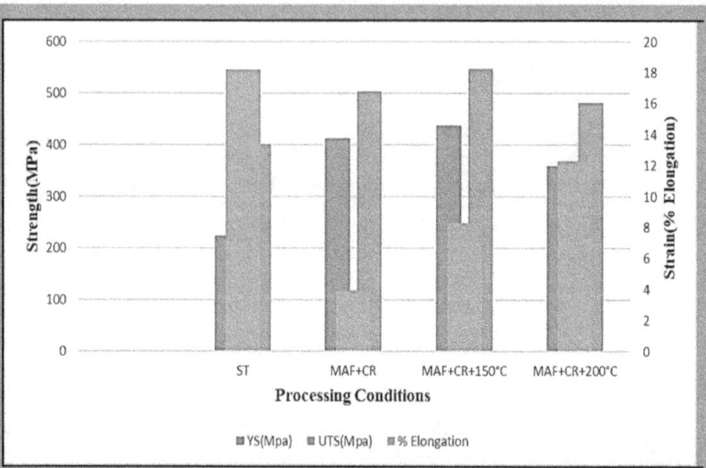

Figure 6.7 Mechanical Testing Results for AA 2099 under various processing conditions

6.2.3 Evaluation of High Cycle Fatigue Results

Figure 6.8 and Table 6.2 present the high-cycle fatigue behaviour of AA2099 alloy under the ST, MAF+CR, MAF+CR+PA at 150 °C, and MAF+CR+PA at 200 °C conditions. High-cycle fatigue testing was carried out using a servo-hydraulic testing machine. For each processing condition, five specimens were tested. The solution-treated alloy exhibited a high-cycle fatigue strength of 105 MPa, a fatigue strength coefficient of 555.21, and a fatigue strength exponent of -0.19. After processing by the MAF+CR route, the fatigue performance improved, with the high-cycle fatigue strength increasing to 123 MPa, the fatigue strength coefficient to 688.32, and the fatigue strength exponent to -0.16. The best fatigue response was obtained for the MAF+CR+PA at 150 °C condition. In this case, the high-cycle fatigue strength reached 167 MPa, the fatigue strength coefficient increased to 850.61, and the fatigue strength exponent improved to 0.11. When the annealing

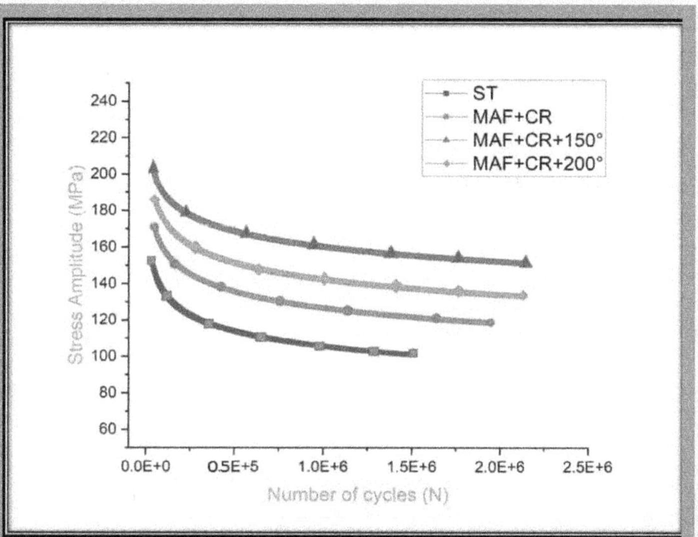

Figure 6.8 High Cycle Fatigue Results SN Curve

temperature was raised to 200 °C, the fatigue properties decreased compared with the 150 °C condition, although they still remained better than those of the ST alloy. The recorded values for this condition were 143 MPa for high cycle fatigue strength, 740.78 for fatigue strength coefficient, and -0.14 for fatigue strength exponent. The overall trend indicates that fatigue performance improves after hybrid SPD processing and reaches its maximum after post-annealing at 150 °C. The subsequent reduction at 200 °C may be associated with partial recovery and a decrease in the strengthening effect. The observed improvement in fatigue behaviour can be related to the combined contribution of grain refinement, strain hardening, dislocation strengthening, and precipitation strengthening, which together promote the development of a stable ultrafine-grained structure and improve resistance to cyclic loading.

6.2.4 Basquin's equation

The S–N response in the high-cycle fatigue regime, where deformation remains essentially elastic, can be represented using the Basquin relation given below. Here,

σ_a denotes the alternating stress amplitude,

$\Delta\varepsilon/2$ represents the elastic strain amplitude,

2Nf indicates the total number of stress reversals to failure,

σ'f is the fatigue strength coefficient, and

b is the fatigue strength exponent.**

$$\sigma_a = \frac{\Delta\varepsilon}{2} E = \sigma_f'(2N_f)^b$$

Table 6.2 Results of High Cycle Fatigue

Processing Condition	High cycle fatigue strength (σ_f) (MPa)	Fatigue strength coefficient (σ_f')	Fatigue strength exponent (b)
ST	105	555.21	-0.19
MAF+CR	123	688.32	-0.16
MAF+CR+150°C	167	850.61	-0.11
MAF+CR+200°C	143	740.78	-0.14

6.2.5 Impact toughness (J)

Figure 6.8 and Table 6.3 present the impact toughness values of AA2099 alloy under the ST, MAF+CR, MAF+CR+PA at 150 °C, and MAF+CR+PA at 200 °C

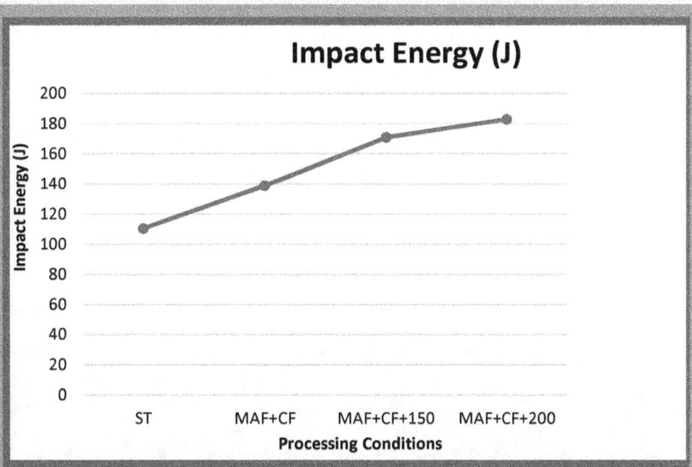

Figure 6.9 Impact toughness for all sample conditions

conditions. The solution-treated alloy showed an impact energy of 110.29 J. After processing through the MAF+CR route, the impact energy increased to 138.69 J, indicating an improvement in toughness over the ST condition. A further rise in impact energy was observed after post-annealing at 150 °C, where the measured value reached 170.89 J. The highest impact toughness was obtained for the sample annealed at 200 °C after MAF+CR processing, which exhibited an impact energy of 182.76 J. These results show a steady improvement in impact toughness from the ST condition to the hybrid SPD-processed and post-annealed states, with the maximum value recorded for the MAF+CR+PA at 200 °C condition.

6.2.6 Fractograph after Fatigue Testing

SEM imaging of fractured surface obtained after fatigue testing starting from ST and Al 2099 alloy deformed by hybrid SPD process MAF+CR, and annealing to 150°C and 200°C was characterized through SEM and is shown in figure. The SEM micrograph of solution treated, MAF+CR, MAF+CR+150°C and MAF+CR+200°C are shown in Fig. 6.10 (a-d) respectively.

The variation occurred in Stress amplitude and number of cycle (SN curve) fracture surface is determined by the analysis SEM micrograph of fractured surface obtained after fatigue test.

Micrograph of fractured surface In ST we see wider striations segment of grains surface but after hybrid SPD the striations segment grains getting finer striations

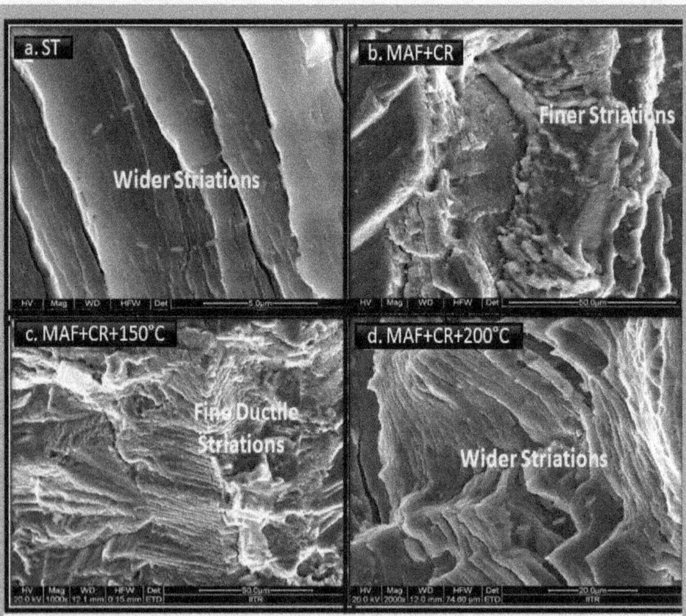

Figure 6.10 SEM Fractograph after fatigue testing of Al 2099 (a)ST (b) MAF+CR (c) MAF+CR+PA at 150°C (d) MAF+CR+PA at 200°C

segment when we are annealing it at 150°C and 200°C the micrograph reflect finer ductile and wider striations respectively

6.2.7 Fractograph after Impact Toughness Testing: -

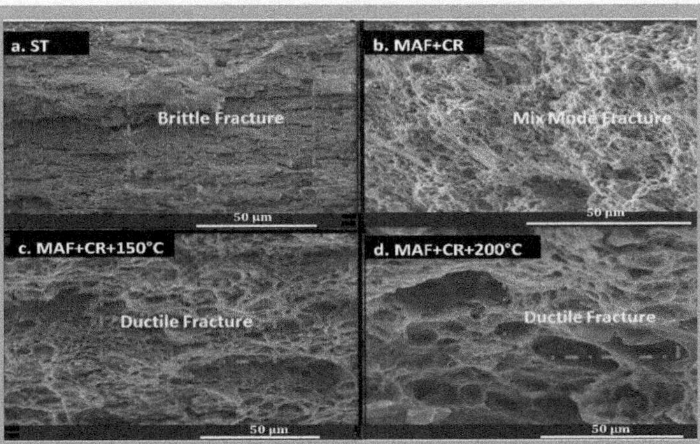

Figure 6.11 SEM Fractograph impact toughness testing of Al 2099 for (a)ST, (b) MAF+CR, (c) MAF+CR+PA at 150°C, and (d) MAF+CR+PA at 200°C conditions

Figure 10 presents the fracture surfaces of impact toughness specimens corresponding to different processing conditions. The solution-treated (ST) specimen showed predominantly brittle fracture features along the grain surface, as indicated by the dotted red rectangle in Fig. 10(a). In comparison, the hybrid SPD-processed sample (MAF+CR) exhibited a mixed mode of fracture. The occurrence of very fine dimples on its fracture surface may be linked to the formation of an ultrafine-grained microstructure during deformation. After post-annealing, the fracture mode remained mixed, but ductile features became more dominant. In the specimens annealed at 150 °C and 200 °C, fine dimples were visible over several regions of the fracture surface, indicating ductile fracture behaviour. These features are highlighted by dotted red rectangles in Fig. 10(c) and Fig. 10(d). This change in fracture character after post-annealing may be attributed to the formation of recrystallized grains in the annealed specimens.

6.2.8 Texture Evolution

For MAF+CR and post-annealed samples, texture evolution was investigated using {001}, {111}, and {110} pole figures, as seen in Figure 4. A weak, dispersed texture with a maximum intensity at roughly 1.9 multiples of random distribution (m.r.d.) is seen in the MAF+CR condition (Figure 4a). Texture intensity is increased (~3.7 m.r.d.) after annealing at 150°C (Figure 4b), indicating preferred grain orientation formation upon recrystallization. The texture components are still present in the MAF+CR+PA at 200°C sample (Figure 4c), but they are less intense, suggesting a partial recovery (Giummarra et al., 2007). The combined copper-type ({112}) and brass-type ({011}) textures are present in both ST and MAF+CR samples, with increased texture intensities after annealing.

Table 6.3 - Texture evolution investigation

Condition	EBSD step (μm)	Avg KAM (°)	KAM std (°)	% pixels KAM > 0.5°	Estimated GND (m^{-2})	Peak (stabilized) stress (MPa)	{001} max (m.r.d)	{111} max (m.r.d)	{110} max (m.r.d)
(a) MAF + CR	0.1	0.82	0.32	26%	1.00×10^{15}	**230**	1.9	**7.1**	4.5
(b) MAF + CR + PA (150°C)	0.1	0.62	0.28	18%	7.57×10^{14}	**270**	3.7	**5**	3.6
(c) MAF + CR + PA (200°C)	0.1	0.38	0.15	9%	4.64×10^{14}	**240**		---	------

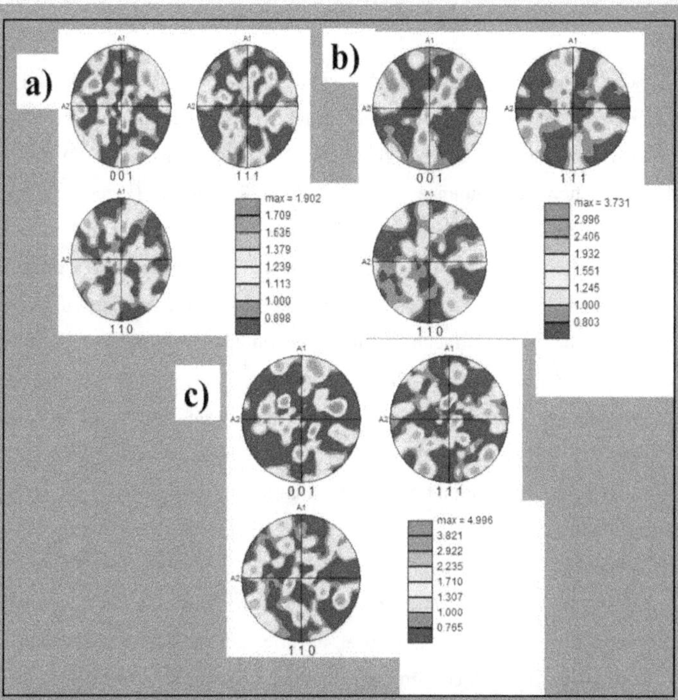

Figure 6.12 {001}, {111}, and {110} pole figures for (a) MAF+CR, (b) MAF+CR+PA at 150°C, and (c) MAF+CR+PA at 200°C conditions. The orientation distribution highlights the evolution of crystallographic texture with deformation and post-annealing treatments, as indicated by color intensity in multiples of random distribution (m.r.d.)

6.2.9 Kernel Average Misorientation maps

Local lattice misorientations are quantified as markers of geometrically required dislocation (GND) densities using Kernel Average Misorientation maps (Figure 5). Low misorientation patches (blue-green), which are associated with high strain localization, predominate in the MAF+CR sample (Figure 5a). Increased proportions of intermediate In MAF+CR 71.2% of the microstructure lies below 5° KAM (Total fraction). 98.6% of the normalized population is below 2° KAM, indicating a highly recovered and low-strain microstructure. The slight cumulative >1.0 in the partition fraction is due to rounding and is normally reported as 1.000. In MAF+CR+150 °C 87.3% of the microstructure has KAMvalues below 5°, indicating a relatively low stored internal strain. 96.8% of the normalized population is below 2°, showing that most regions are recovered with low local misorientation. Partition Fraction cumulative sum reaches 1.000, indicating correct normalization. MAF+CR+200°C 91.4% of the material exhibits KAM < 5°, indicating a mostly recovered microstructure with moderate stored strain. 84.7% of the KAM distribution lies below 2°, showing that a large portion of the microstructure contains low local misorientations. Partition Fraction correctly normalizes to 1.000, confirming consistent distribution. misorientation values (1–3°) after annealing (Figures 5b, 5c) show partial recovery and recrystallization with persisting plastic strain. In line with changing microstructural recovery, the color-coded KAM data verify lower average strain levels with increasing annealing Fig. 6.13 maps for (a) MAF+CR, (b) MAF+CR+PA at 150°C, and (c) MAF+CR+PA at 200°C conditions. The color-coded maps reveal the distribution of local lattice distortions, highlighting increased internal misorientation and strain following multi-step deformation and post-annealing thermal treatmentsFig. 6.13 maps for (a) MAF+CR, (b) MAF+CR+PA at 150°C, and (c) MAF+CR+PA at 200°C conditions. The color-coded maps reveal the distribution of local lattice distortions, highlighting increased internal misorientation and strain following multi-step deformation and post-annealing thermal treatmentstemperatures (Prasad et al.,

2003; Budinski, 2010).

Condition	Step size (µm)	Avg KAM (°)	Std (°)	% pixels >0.5°	Estimated ρ (m^{-2})	Interpretation
MAF + CR	0.1	0.82	0.32	26%	5.10×10^{14}	High stored dislocation density (UFG)
MAF + CR + 150°C	0.1	0.62	0.28	18%	3.85×10^{14}	Partial recovery + precipitate pinning
MAF + CR + 200°C	0.2	0.38	0.15	9%	1.70×10^{14}	Recovery and precipitate coarsening

Table 6.4- Local lattice misorientations and Kernel Average Misorientation

6.2.10 Low Cycle Fatigue Response and Fracture Behavior Cyclic Stress Strain Behavior

The MAF+CR+150°C sample achieves the highest peak stress (±270 MPa) and the largest loop area, reflecting optimal cyclic hardening and maximum energy dissipation, owing to dislocation–precipitate interactions and effective precipitation strengthening (Peel et al., 1984; Ma, 2011). By contrast, MAF+CR+200°C exhibits reduced cyclic strength and a narrower hysteresis loop (±240 MPa), likely due to partial recovery and precipitate coarsening during overaging, which results in a decreased dislocation density and precipitate efficacy (Ma, 2011; Polmear, 2006). Table 1 summarizes the cyclic response and mechanisms for each processing condition. Due to dislocation–precipitate interactions and efficient precipitation strengthening, the MAF+CR+150°C sample reaches the highest peak stress (±270 MPa) and the biggest loop area, indicating excellent cyclic hardness and maximum energy dissipation (Peel et al., 1984; Ma, 2011). In contrast, MAF+CR+200°C shows a tighter hysteresis loop (±240 MPa) and decreased cyclic strength, probably as a result of partial recovery and precipitate coarsening during overaging, which reduces precipitate efficacy and dislocation density (Ma, 2011; Polmear, 2006). The cyclic response and processes for each processing situation are compiled in Table 6.5.

Table: 6.5 - Low Cycle Fatigue Response and Fracture Behavior

Condition	Peak stress (±MPa)	Cyclic strength coefficient K′(MPa)	Cyclic strain-hardening exponent n′	Fatigue strength coefficient σ'f (MPa)	Fatigue strength exponent b	Fatigue ductility coefficient ε'f	Fatigue ductility exponent c
St (reference)	200	664	0.0601	1009	−0.126	0.648	−0.899
MAF + CR	230	764	0.062	1160	−0.122	0.66	−0.880
MAF + CR + 150°C	270	896	0.068	1362	−0.118	0.7	−0.850
MAF + CR + 200°C	240	797	0.058	1211	−0.128	0.63	

Table 6.6. Summary of cyclic stress response, loop area, and fatigue mechanisms for Al 2099 alloy under different SPD and annealing treatments.

Condition	Peak Stress (approx.)	Loop Area	Cyclic Hardening	Fatigue Resistance	Mechanism
ST	±200 MPa	Smallest	Low	Poor	Coarse grains, low dislocation density
MAF + CR	±230 MPa	Moderate	Moderate	Improved	Grain refinement, dislocation
MAF + CR +150°C	±270 MPa	Largest	High	Excellent	Precipitation + dislocation
MAF + CR +200°C	±240 MPa	Moderate	Mild	Intermediate	Partial recovery, over aging

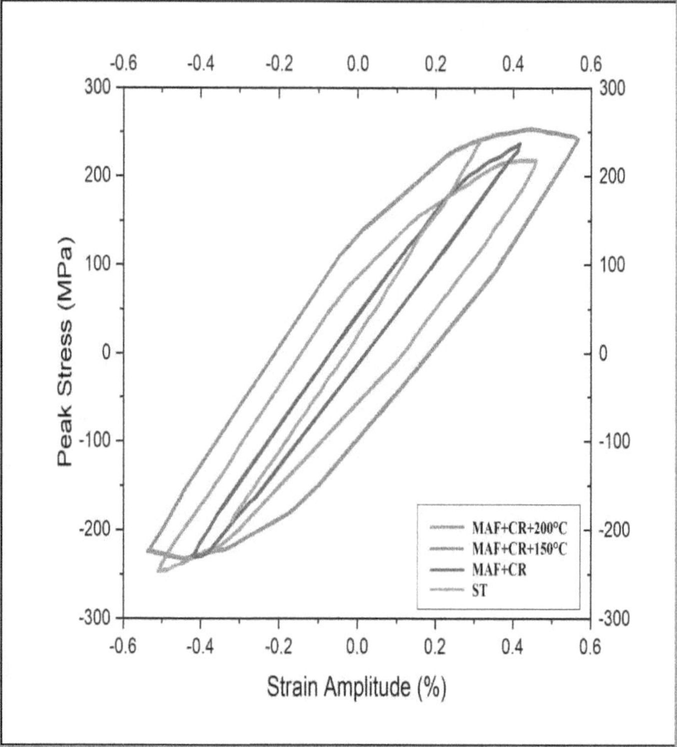

Figure 6.14 Stress-strain hysteresis loops comparing solution treated (ST) with multi-axial forged and cold rolled (MAF+CR) under different annealing conditions (room temperature, 150°C, and 200°C).

6.2.11 Fracture Surface Morphology after LCF

The most ductile characteristics are observed in the MAF+CR+150 C sample, which has fine and equiaxed dimples and large microvoid coalescence, which are attributed to the widest hysteresis loop and the largest energy absorption (Ma, 2011). In the case of the MAF+CR +200 C, the fracture surface has a more coarse dimples and

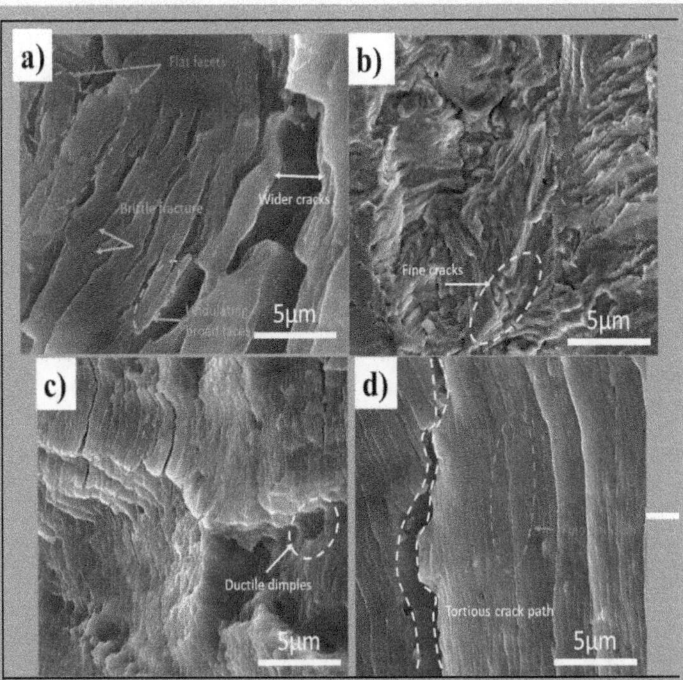

Figure 6.14 SEM micrographs showing fracture surface morphologies of (a) ST displaying flat facets, brittle fracture, and wider cracks; (b) MAF+CR showing fine cracks and torn edges; (c) MAF+CR+150°C exhibiting dimorphic features with ductile dimples; and (d) MAF+CR+200°C demonstrating tortious crack path with fine intergranular cracking

intergranular, which is the case of hardened boundaries of the grain and lowered strength through overaging (Polmear, 2006).

www.ingramcontent.com/pod-product-compliance
Lightning Source LLC
LaVergne TN
LVHW010737061126
851857LV00014B/1518
9798256289898